T0143376

5G Outlook – Innovations and Applications

RIVER PUBLISHERS SERIES IN COMMUNICATIONS

Volume 48

The "River Publishers Series in Communications" is a series of comprehensive academic and professional books which focus on communication and network systems. The series focuses on topics ranging from the theory and use of systems involving all terminals, computers, and information processors; wired and wireless networks; and network layouts, protocols, architectures, and implementations. Furthermore, developments toward new market demands in systems, products, and technologies such as personal communications services, multimedia systems, enterprise networks, and optical communications systems are also covered.

Books published in the series include research monographs, edited volumes, handbooks and textbooks. The books provide professionals, researchers, educators, and advanced students in the field with an invaluable insight into the latest research and developments.

Topics covered in the series include, but are by no means restricted to the following:

- Wireless Communications
- Networks
- Security
- Antennas & Propagation
- Microwaves
- Software Defined Radio

For a list of other books in this series, visit www.riverpublishers.com

5G Outlook – Innovations and Applications

Editor

Ramjee Prasad

Founder Director
Center for TeleInFrastruktur (CTIF)
Aalborg University
Denmark

Founder Chairman
Global ICT Standardization Forum for India (GISFI)
India

Published, sold and distributed by:
River Publishers
Alsbjergvej 10
9260 Gistrup
Denmark

River Publishers
Lange Geer 44
2611 PW Delft
The Netherlands

Tel.: +45369953197
www.riverpublishers.com

ISBN: 978-87-93379-77-0 (Hardback)
978-87-93379-78-7 (Ebook)

Dedicated to
My Colleagues from the CTIF Global Network

Contents

Preface

आत्मौपम्येन सर्वत्र समं पश्यति योऽर्जुन
सुखं वा यदि वा दुःखं स योगी परमो मतः ॥ ३ २ ॥

Transliteration

ātmaupamyena sarvatra samaṁ paś yati yo 'rjuna
sukhaṁ vā yadi vā duhkhaṁ sa yogī paramo matah

Translation

One who perceives in comparison with the self, all living entities equally, in happiness and distress; such a one perfected in the science of uniting the individual consciousness with the ultimate consciousness in considered the highest.

The Bhagavad Gita (6.27)

This book is outcome of contributions by the researchers from several countries coming from universities, research institutes and industries. The topics covered in this book vary from basic concept of millimeter-waves and their characteristics and usages in 5G mobile technology to the real-time and future applications of 5G in humans' daily life and work. Details on technical issues of 5G wireless networks such as effects of denial of service on the network layer and positioning of devices with 5G using flexible wavelet packet modulation are addressed in this book. The chapters of this book are part of the 5th IEEE 5G Summit held in Aalborg on the first July 2016.

IEEE 5G summit is a one-day summit, held on 26 May 2015 at Princeton University, New Jersey for the first time, co-sponsored by Center for TeleInfrastruktur-United States of America (CTIF-USA). The event is organized by IEEE Communications Society. The main purpose for this event is engaging the industry and university with high value and innovative technologies in 5G and its features. The platform provided by 5G summit event is supposed to connect industry leaders, innovators and researchers from academic area to exchange ideas in the divers areas related to 5G and future plans for wireless technology. IEEE 5G summit is planned to hold a series in different countries. Princeton 5G summit was followed by Toronto 5G Summit on November 14, 2015 and Silicon Valley 5G summit on November 16, 2015. There was the first 5G summit event in Asia on March 29, 2016 at Indian Institute of Technology (IIT) Patna, India jointly with the 24th *GISFI* Standardization Series Meeting (GSSM). This book is collection of some presentations at 5G summit event held in Aalborg (first one in Europe) on July 1, 2016. IEEE Communications Society together with the steering Committee of 5G summit event are planning more events in future to exchange the latest findings in 5G areas and proposing the proper solutions for the future of wireless technology.

5G stands for fifth generation mobile technology. The 5G is the future of the communication technology after 1G, 2G, 3G and 4G. Both industries and universities found that higher data rate, more reliable communications and better Quality of Service (QoS) should be achieved within 5G technology. The journey of mobile generations has started from 1980s and each generation completed the previous one's lacks. For example Global System for Mobile Communications (GSM) presented after 1G to fix the security weakness of analog communication systems. 3G was introduced to cover GSM's lack of mobile data; however it was not a huge success. Then 4G was needed to make better data rate. Now 5G is presented to continue this journey to reach the peak of wireless technology. There have been lots of research and works on 5G from coding and modulation techniques, advanced signal processing, multi-hop networks until applications and standardizations. But there are still a huge amount of works, research and implementations to do. It is predicted that the 5G concept becomes a reality by 2020.

This book consists of 14 chapters covering different challenges and innovations in 5G technology. The book starts with a comprehensive Introduction about 5G from history to the future and continues with the second chapter including the concept of 5G, following by millimeter-waves and their promises to 5G. Two new ideas are presented in Chapters 4 and 5. The book introduces a history and future of International Mobile Telecommunications (IMT) in Chapter 6 and follows detailed technical issues in the signal processing and network design for 5G in two upcoming chapters. The rest of the book is dedicated to the diverse applications of 5G in different areas from health science to the business modeling.

Ramjee Prasad
Founder Director, CTIF, Aalborg University, Denmark
Founder chairman, GISFI, India

Acknowledgments

This book is the collection of diverse works and research done by different colleagues worldwide for the 5th IEEE 5G Summit held on July 1, 2016 in Aalborg. Therefore, the editor would like to thank all the contributors involved in this book for their collaboration and dedication that made this book finalized. The editor would also like to thank all organizer and participants in this event for sharing and exchanging their findings and novel ideas for the future of 5G technology.

The editor would like to acknowledge the contributions from Center for TeleInFrastrukture (CTIF) colleagues for their role in initiating 5G research at CTIF namely, Albena, Neeli, Sofoklis, Ambuj, Prateek, Mihail, and so on.

Finally, the editor likes to express his special thanks to Maryam for her effort on formatting this book.

About the Editor

Ramjee Prasad is the Founder President of the CTIF Global Capsule (CGC). He has been a Founding Director of the Center for TeleInFrastruktur (CTIF) since 2004. He is also the Founder Chairman of the Global ICT Standardisation Forum for India (GISFI: www.gisfi.org) established in 2009. GISFI has the purpose of increasing of the collaboration between European, Indian, Japanese, North-American and other worldwide standardization activities in the area of Information and Communication Technology (ICT) and related application areas.

He was the Founder Chairman of the HERMES Partnership – a network of leading independent European research centres established in 1997, of which he is now the Honorary Chair. He is a Fellow of the Institute of Electrical and Electronic Engineers (IEEE), USA, the Institution of Electronics and Telecommunications Engineers (IETE), India, the Institution of Engineering and Technology (IET), UK, Wireless World Research Forum (WWRF) and a member of the Netherlands Electronics and Radio Society (NERG), and the Danish Engineering Society (IDA).

He has received Ridderkorset af Dannebrogordenen (Knight of the Dannebrog) in 2010 from the Danish Queen for the internationalization of top-class telecommunication research and education. He has been honored by the University of Rome "Tor Vergata", Italy as a Distinguished Professor of the Department of Clinical Sciences and Translational Medicine on March 15, 2016.

He has received several international awards such as: IEEE Communications Society Wireless Communications Technical Committee Recognition

Award in 2003 for making contribution in the field of "Personal, Wireless and Mobile Systems and Networks", Telenor's Research Award in 2005 for impressive merits, both academic and organizational within the field of wireless and personal communication, 2014 IEEE AESS Outstanding Organizational Leadership Award for: "Organizational Leadership in developing and globalizing the CTIF (Center for TeleInFrastruktur) Research Network", and so on.

He is the Founder Editor-in-Chief of the Springer International Journal on Wireless Personal Communications. He is a member of the editorial board of other renowned international journals including those of River Publishers. Ramjee Prasad is Founder Co-Chair of the Steering committees of many renowned annual international conferences, e.g., Wireless Personal Multimedia Communications Symposium (WPMC); Wireless VITAE and Global Wireless Summit (GWS).

He has published more than 30 books, 1000 plus journals and conferences publications, more than 15 patents, over 100 Ph.D. Graduates and larger number of Masters (over 250). Several of his students are today worldwide telecommunication leaders themselves.

List of Figures

List of Tables

List of Abbreviations

AAS	Active Antenna System
AMPS	Advanced Mobile Phone System
AOA	Angle of Arrival
AWGN	Additive White Gaussian Noise
BCC	Body Channel Communication
BMES	Business Model Ecosystems
BMI	Business Model Innovation
BSS	Basic Service Set
BTS	Base Transceiver Station
BYOD	Bring Your Own Device
CAM	Computer Aided Manufacturing
COTS	Commercial Off-The-Shelve
CR	Cognitive Radio
DCM	Dual Carrier Modulation
DoS	Denial of Service
DSA	Dynamic Spectrum Access
FBMC	Filter Bank Multi-Carrier
FIR	Finite-Impulse-Response
FM	Frequency Modulation
FPLMTS	Future Public Land Mobile Telecommunications Systems
GISFI	Global ICT Standardization Forum for India
GPRS	General Packet Radio Services
GPS	Global Positioning System
HBC	Human Bond Communications
HMI	Human to Machine Interface
HPF	High Pass Filter
HSPA	High Speed Packet Access
ICC	International Conference on Communications
ICT	Information and Communication Technology
IDMA	Interleave Division Multiple Access
IFTTT	If This Then That

IMT	International Mobile Telecommunications
IoT	Internet of Things
IP	Internet Protocol
ISI	Inter Symbol Interference
ISSE	International Symposium on Systems Engineering
ITU	International Telecommunications Union
LBT	Listen before Talk
LDS	Low Density Spreading
LOS	Line of Sight
LPF	Low Pass Filter
LTE	Long Term Evolution
MAC	Medium Access Control
MIMO	Multiple Input Multiple Output
MM	Mixed Mode
MSE	Mean Square Error
MU	Multi User
NFV	Network Function Virtualization
NLOS	Non Line of Sight
NMT	Nordic Mobile Telephone
NOC	Network Operation Center
OFDM	Orthogonal Frequency Division Multiplexing
OTT	Over The Top
PBM	Persuasive Business Models
PCS	Personal Coaching System
PDMA	Pattern Division Multiple Access
PPM	Pulse Position Modulation
PU	Primary User
PUEA	Primary User Emulation Attack
QMF	Quadrature Mirror Filter
QoS	Quality of Service
RAN	Radio Access Network
RAT	Radio Access Technologies
RSMT	Remotely Supervised Myofeedback Treatment
RSS	Received Signal Strength
RU	Resource Unit
SCMA	Sparse Code Multiple Access
SDN	Software Defined Networking

SDP	Semi Definite Programming
SINR	Signal to Interference Noise Ratio
SMNAT	Smart Mobile Network Access Topology
SMS	Short Message Service
SNR	Signal to Noise Ratio
SON	Self Organizing Network
SU	Secondary Users
TACS	Total Access Communication Systems
TDOA	Time Difference of Arrival
TH	Time Hopping
TOA	Time of Arrival
UWB	Ultra Wide Band
VLC	Visible Light Communications
VTC	Vehicular Technology Conference
WPM	Wavelet Packet Modulated
WRC	World Radio Communication Conferences

1

Introduction

Maryam Rahimi and Ramjee Prasad

Center for TeleInfrastruktur (CTIF), Aalborg University, Denmark

Radio technologies started their first step to connect the whole world together by introducing the analogue cellular systems in 1980s. Afterwards radio technologies have experienced a rapid evolution in their paths with the launch of the digital wireless communication systems in different steps from Global System for Mobile Communications (GSM) to the Fifth Generation Mobile Technology (5G). 5G technology promises to evolve the mobile communication system with its features especially within very high bandwidth. It is predicted, users experience a huge difference between 5G technology and previous mobile generations, which makes 5G technology more powerful in near future.

5G promises its users enjoying high speed up to 1Tera bps (1Tbits/s = 10^{12} bits/s) wireless link rates, wider range of applications, connectivity everywhere, watching high quality videos on their cell phones, lower battery consumption and much more around the year 2020, where it is promised to implement 5G in the reality. For now still huge amount of works and researches are needed to configure 5G standards, network design, signal processing and many other issues to be addressed.

1.1 The Journey to 5G Wireless Communication

The Journey began in 1819, when the Danish physicist Chritian Oersted founded the fundamental relationship between electricity and magnetism. Nowadays we know it as electromagnetic field. In 1832, Michael Faraday used the concept of electromagnetic and discovered a way to provide electricity.

Later on James Clerck Maxwell introduced his magic equations in 1865 and 1873. The fundamental of critical inventions in next decades are defined by Maxwells' equations.

The road towards radio communication started in fact in 1895, when Guglielmo Marconi sent radio telegraph transmission across the English Channel. Similar works by other contemporary scientist, Jagdish Chandra Bose from India pioneered several dimensions towards the field of radio communications. The journey continued with first voice over radio transmission in 1914 and first long-distance TV transmission in the United States, conducted by AT&T Bell Labs in 1927. Frequency modulation (FM) was introduced by Armstrong in 1935, which was a huge step forward at that time.

In the 1980s, the First Generation Mobile Networks (1G) were introduced. The 1G signaling systems were designed based on analog system transmissions. Some of the most popular standards set up for 1G system were Advanced Mobile Phone System (AMPS), Total Access Communication Systems (TACS) and Nordic Mobile Telephone (NMT).

In the early 1990s, the Second Generation Mobile Networks (2G) based on Global System for Mobile Communications (GSM) was launched. 2G is the starting point towards wireless digital communication. The main concern of new system was covering the problems with security weakness of analog communication systems. Another novelty in designing GSM was using digital modulation to improve voice quality but the network offers limited data service. The 2G carriers also began to offer additional services, such as paging, faxes, text messages and voicemail.

An intermediary phase, 2.5G was introduced in the late 1990s. It uses the General Packet Radio Services (GPRS) standard, which delivers packet-switched data capabilities to existing GSM networks. It allows users to send graphics-rich data as packets. The importance for packet-switching increased with the rise of the Internet and the Internet Protocol (IP).

The Third Generation Mobile Systems (3G) is proposed in 2000s to provide high speed Internet access to allow mobile phone customers using video and audio applications. One of the main objectives of designing 3G system was to standardize on a single global network protocol instead of the different standards adopted previously in Europe, the U.S. and other regions. 3G phone speeds deliver up to 2 Mpbs, but only under the best conditions and in stationary mode. Moving at a high speed can drop 3G bandwidth to a mere 145 Kbps [1].

The Fourth Generation Mobile System (4G) offered in 2010s promised for providing transmission rates up to 20 Mbps. The concept of Quality of Service (QoS) got more attention in 4G system. It is promised accommodating QoS features in 4G system. QoS will allow telephone carrier to prioritize traffic according to the type of application using bandwidth and adjust between different telephones needs at a moment's notice. High quality video and audio streaming over end to end Internet Protocol is the most attraction of 4G. There are two important standards in 4G technologies; Worldwide Interoperability for Microwave Access (WiMax) and Long Term Evolution (LTE). 4G is currently using in many countries all over the world [2]. Figure 1.1 shows a trend in cell phones and their development to the future.

The journey is continuing to the Fifth Generation Mobile System (5G) to achieve the huge promises made by 5G in terms of high data rate, low latency, low power consumption and much more. It is a big step forwards during this road and no one knows if this step makes the human to achieve the peak of the wireless communications or still there is a long way ahead.

Figure 1.1 Cell phone development to the future.

1.2 Background and Future of 5G Technology

The concept of realizing 5G communication network based on Wireless Innovative System for Dynamic Operation Mega Communication (WISDOM) followed by other leading initiatives at research facilities in industry and academia is as shown in Table 1.1 [3].

The future of 5G technology is under discussion and research. It is expected that 5G technology increases the bandwidth, QoS, improves usability and security, decreases delays and cost of services, reduces battery consumption, improves reliability of the communications and much more. It is mentioned that, the architecture of 5G is highly advanced and service providers can implement the new technology easily [6].

The challenges of 5G such as, Inter-cell interference, traffic management, multiple services, security and privacy, standardization and so on should be addressed before implementation of 5G technology, which is expected for 2020. It needs lots of efforts from universities and industries to exchange their findings and ideas to reach the final and promised point in 5G system.

Table 1.1 Significant 5G initiatives till date

Year	5G Initiative	Entity	Country
2008 February	WISDOM [3]	Center of TeleInFrastruktur, Aalborg University	Denmark
2008 November	5G through WISDOM [4]	Center of TeleInFrastruktur, Aalborg University	Denmark
2008 November	5G systems based on Beam Division Multiple Access [4]	South Korea IT R&D department	South Korea
2012 May	First 5G System [4]	Samsung Electronics	South Korea
2012 October	5G Research Center [4]	University of Surrey	United Kingdom
2013 November	Research on 5G systems [4]	Huawei Technologies Co. Ltd	China
2014	5G: 2020 and Beyond [5]	Center of TeleInFrastruktur, Aalborg University	Denmark

1.3 Applications of 5G

Researchers expected 5G technology supports the speed up to 1 Tbit/s, less than 1ms latency, almost 10% network coverage, 1000 times reduction in power consumption, deep indoor coverage, 10 to 100 connected devices, 100x average data rate and so on. The question is what users can do if all promises will be met by 5G technology? Actually the applications of 5G is very much equivalent to achievement of dream. Users could expect unified global standard, network availability anywhere any time, Wireless Fidelity (Wi Fi) global zone and much more. With this great offer from 5G wireless technology people are able to play online games against friends across the world with lots of new experiences, cars can drive automatically, which is a big help to transportation system, health care could be done remotely, a great help to elderly people and video conferencing becomes much more real like you are at the same room with a person who you talk to.

One of the most inspiring application of 5G is Human Bond Communication, proposed in 2015 [7]. Human bond communication is a novel concept that incorporates olfactory, gustatory, and tactile that will allow more expressive and holistic sensory information exchange through communication techniques for more human sentiment centric communication. 5G is the tool to achieve this idea as long as it supports very high data rates. Figure 1.2 shows how the concept of Human bond communication could work in the reality.

Visible Light Communications (VLC) is another useful application of 5G for ambient assisted living discussed in 2014 [8]. VLC is an alternative to communication technology using RF wireless spectrum and proposes a concept for integrating VLC to enable intelligent communication infrastructure. The VLC systems can be deployed along with a repeater system for targeted discrete deployment as shown in Figure 1.3.

5G technology can evolve wireless health monitoring system due to its speed, reliability, security and so on. Elderly monitoring system could develop to remote surgery. It is a huge help to people from not very developed countries to receive the same care and medical services as other people. 5G technology will promises us better future in the world more equality and more welfare.

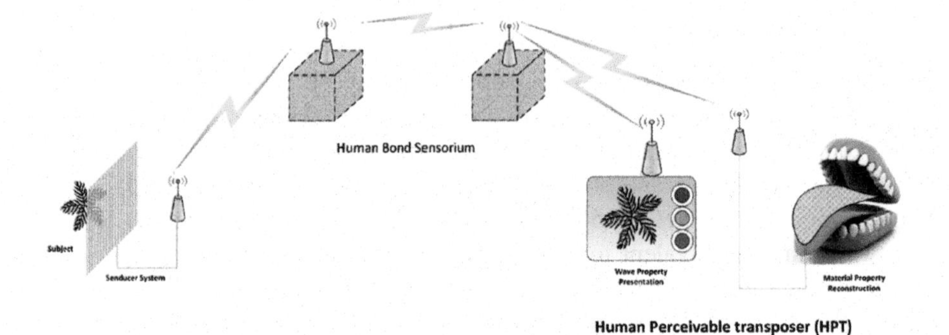

Figure 1.2 Human bond communication in real world.

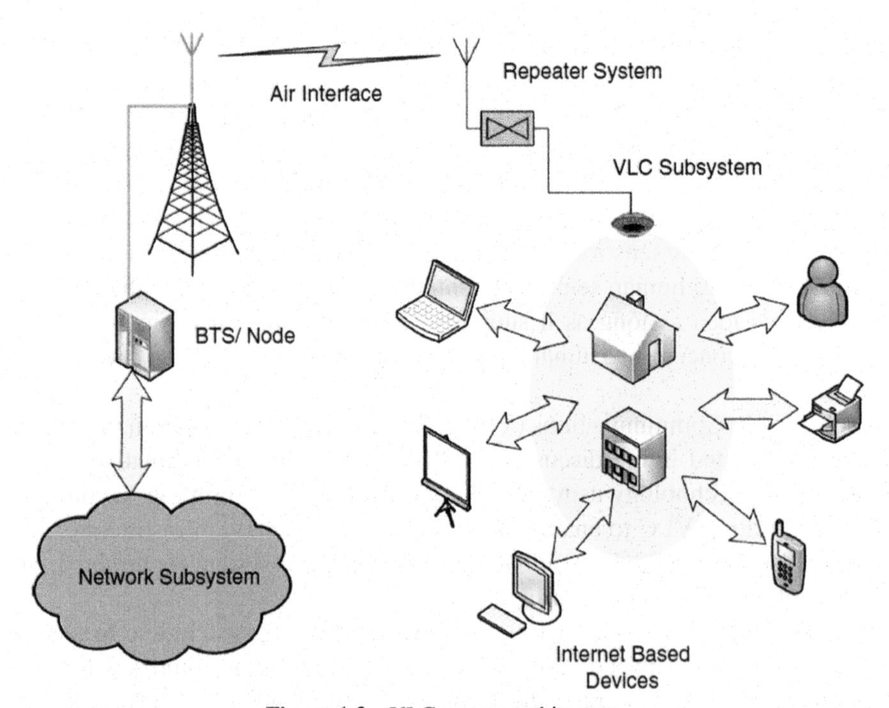

Figure 1.3 VLC system architecture.

1.4 Summary

Mobile wireless communications do not stop growing; the evolution from 1G to 4G has developed the quality of human life dramatically. Lots of dreams came to reality by wireless communication technologies and applications.

5G as the next generation of mobile communication promises to evolve the wireless technology with its speed, reliability and other features. It is expected that, implementation of 5G happens in 2020. Till that time researchers and industry leaders need to carry on different works and exchange data and ideas to achieve the 5G promises. This book presents the latest research results and innovative ideas from both university and industry.

References

[1] Ramjee Prasad, Werner Mohr and Walter Konhäuser, "Third Generation Mobile Communication Systems", Artech House, 2000.

[2] Shinsuke Hara and Ramjee Prasad, "Multicarrier Techniques for 4G Mobile Communications", Artech House, 2003.

[3] Ramjee Prasad, Keynote Speech – Wireless Innovative System Dynamic Mega Communications (WISDOM), in IEEE CogART 2008, Feb., First IEEE International Workshop on Cognitive Radio and Advanced Spectrum.

[4] 5th Generation Mobile Networks, Wikipedia [online], http://en.wikipedia.org/wiki/5G

[5] Ramjee Prasad, "5G: 2020 and Beyond", River Publisher, 2014.

[6] Ramjee Prasad, "Global ICT Standadization Forum for India (GISFI) and 5G Standardization", Journal of ICT Standardization, Vol. 1, No. 2, pp. 123–136, Nov. 2013.

[7] Ramjee Prasad, "Human Bond Communication", Wireless Personal Communication, Vol. 87, No. 3, pp. 619–627, 2015.

[8] Ambuj Kumar, Albena Mihovska, Sofoklis Kyriazakos and Ramjee Prasad, "Visible Light Communications (VLC) for Ambient Assisted Living", Wireless Personal Communication, Vol. 78, No. 3, pp. 1699–1717, 2014.

About the Author

Maryam Rahimi is currently a Research Assistant at CTIF, Aalborg University. She has studied her Ph.D. at Aalborg University in Wireless Communications and her Master at University Putra Malaysia in Microelectronics. Her focus during Her Ph.D. was at MIMO, signal processing and channel propagation. Maryam has some industrial work experience as an Electronics Engineer. She was involved in different projects and research namely, Design and Development of Radio Frequency (RF) Front-end for a Wireless Receiver or Transceiver, project of ministry science, 2008–2009 and Analysis of Measured Channels, Channel Propagation, Cooperative MIMO Capacity, 2010–2014. She is author and co-author of more than 15 Journal and Conference papers.

2

5G: Need for the Hour

Nidhi[1] and Ramjee Prasad[2]

[1]Vishwaniketan, Navi Mumbai, India
[2]Center for TeleInfrastruktur (CTIF), Aalborg University, Denmark

2.1 Introduction

Current legacy mobile networks in lace, offer plethora of services. Mobile technology is the most ubiquitous technology and has influenced the society more than any other technology. The society has witnessed technological advancements in past years which have dramatically changed the way mobile and wireless communication systems are being used. The appetite for broadband has clearly fuelled the development of mobile cellular networks.

The revolution in mobile communication had always surprised mankind. Within the time span of almost a decade, mobile and wireless communication technology has something big to give. Each succeeding generation has successfully fixed the loopholes of its predecessor and had rooms for the next generation. The 5G is the future of the communication after the success of 1G, 2G, 3G and the recent 4G. 5G is still under developed and fighting for standards which will exist in 2030s and beyond.

Today, almost every telecom company is pulling their socks up to bring 5G standards. 5G networks are considered to be the next generation's "Smart" networks. It will incorporate intelligence with speed which will help management of billions of connected devices and numerous emerging technologies. 5G networks will emerge as an enabler for connected world. 5G technology aims to enhance the connectivity and mobile broadband. The swift from 3G to 4G has witnessed vast developments. The driving force behind 4G was Speed and it introduced fast video streaming with fast mobile broad bandwidth.

5G will not only enhance the speed but upgrade the technology expansively [1]. Automation services, service-aware networks, context awareness, real-time monitoring services, critical services and many more application areas will dominate the society. 5G networks will introduce new levels intelligence both at device and at network level followed by various emerging disruptive technologies. It will enable devices and network elements to automatically sense the requirements, situation, availability and so on, based on which it accordingly adjust connection speed, degree of latency, alert messages, device power configurations and so on.

The eminent 5G features are broadly higher bandwidths, lower latencies, several times faster data rates, highest degree of scalability, high throughputs, enhanced capacity and so on. It provides opportunities to different technologies which can be exploited to achieve desired level of configurations to meet 5G standards. 5G will be used to meet the requirements which were untethered with the communication generations familiar to us. The application areas and potential 5G target includes zero-outage probability, connecting rural and remote areas, universal internet connectivity, virtual reality, real-time applications and so on.

5G will dominate other mobile networks if it's said standards and features are met. The initiatives and research on 5G is burgeoning at an alarming rate. The race is just not only between the telecom companies now, social media companies like Google and Facebook are also competing for 5G. The race is ramping up to reach gigabit speeds per second, increased capacity and mainly rural connectivity at an economic and affordable rate.

The figures are shocking but there is still a huge percentage of population that is deprived of even basic connectivity. The technologies have grown and developed better in urban scenarios but the rural population is left uncovered from the radar of connectivity. Thus, 5G aims to reduce this gap and the rural and sparsely populated regions to provide basic connectivity along with seamless internet connectivity.

2.2 Mobile Communication Aeon

The wireless mobile communication technologies have gradually evolved to serve the society and meet its basic needs. The technology was analogue when it first came into picture, then came the era of digitization which left its remarkable impact on communication systems and today we have advanced digitized technologies with additional features. There always have been a driving force for the development and deployment of any technology, whether it is analog or the recent evolving "Long Term Evolution (LTE)" and "LTE-Advanced (LTE-A)".

The need for untethered telephony with wireless real-time voice communication has dominated the success of cordless phones, followed by the "First Generation of Mobile Communication (1G)". The transition from 1G to "Second Generation of Mobile Communication (2G)" characterised "Short Message Service (SMS)" text messaging as its killer application. Gradually with the widespread usage of computers, Internet ruled the technological advancements and filed its candidature as the killer application for the next successor which was "Third Generation of Mobile Communication (3G)" [2]. The emergence of smartphones that integrated the features of computers with the cellular technology to provide services at fingertips and eventually gave birth to the "Fourth Generation of Mobile Communication (4G)", famously known as LTE. The access to faster data speeds and video streaming were the killer applications in 4G. The 4G evolved with all IP services as its key differentiator but raised the unanswered question for power efficiency and low frequency networks in densely deployed environment [3].

With "Fifth Generation of Mobile Communication (5G)", mobile operators would create a society offering massive connectivity which will act as an enabler for Machine-to-Machine (M2M) services and the Internet of Things (IoT).

2.3 WISDOM and Its Task Groups Abstract

"Wireless Innovative System for Dynamically Operating Mega Communications (WISDOM)" is the standard for the 5G mobile communication to be developed in India by 2020. It is a novel technology based on the Cognitive Radio which employs the feature of the opportunistic sensing of the spectrum holes or white spaces. The state of Art technology generations which have already been implemented or are being implemented are 1G, 2G, 3G and 4G. The 4G technology being the latest in the mobile communications serves to provide 100 Mbps of data services. But still the demand for services has not seen saturation. Therefore there is a need to forge a new generation of Mobile Communication which can take care of arising services like wearable or flexible mobile devices, UHD video streaming, navigation and cloud services and many more. WISDOM is proposed to be providing 1Tbits/s of speed at Mm-wave frequencies at short range communication and 300 Mbps of data rate for individual mobile subscribers.

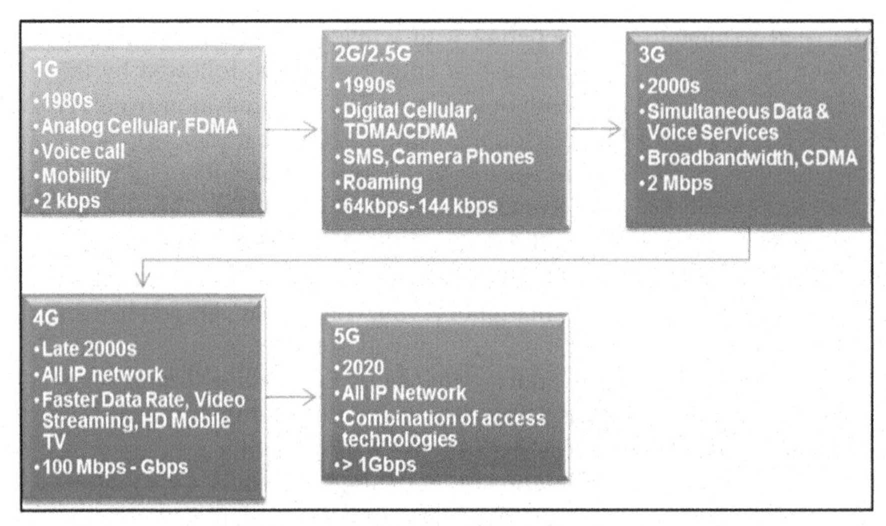

Figure 2.1 Mobile wireless technology.

"Global ICT Standardization Forum for India (GISFI)" is an organization is involved with the standardization of the 5G standard [4]. For the standardization it has five existing task groups namely Security and Privacy, Future Radio Networks and 5G, Internet of Things, Cloud and Service Oriented Network and Green ICT. Recently two new task groups are added namely Spectrum Group and Special Interest Group. Each of the above groups has their own tasks and WISDOM presents challenges in each of the task groups. The challenges are to be sequentially met with one task group completing the challenge the other task group takes the next challenge. The Figure 2.2 suggests the pathway to 5G technology. The concept of Collaborative Communication, Navigation, Sensing and Services (CONASENSE) and Human Bond Communication (HBC) elaborates the idea for 5G together with WISDOM. CONASENSE suggests that all the operations will be carried out collectively and the information will be shared between the elements as they will inter relate among themselves to give high end services to the users [5] while HBC emphasises that with the increase in bandwidth in the successive generations there exists a window where we can look after the other three senses which are touch, smell and taste [6]. Thus, there are challenges needs to be taken care in the 5G.

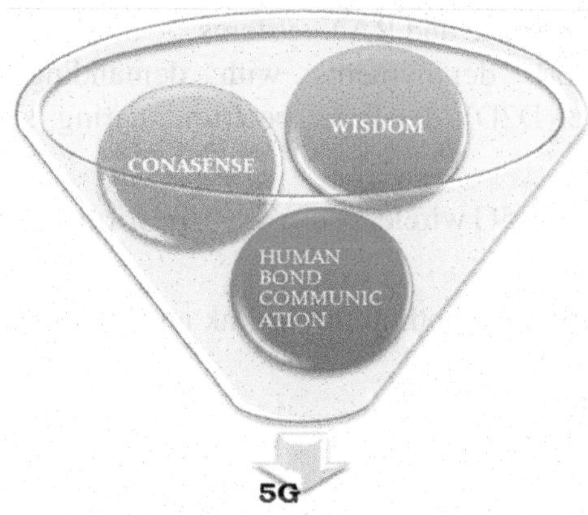

Figure 2.2 Collective CONASENSE, WISDOM & HBC for 5G.

2.4 Towards 5G System

This Section discusses about need for 5G technology as well as different applications such as use cases, self-driven cars and so on.

2.4.1 Requirements and Drivers

One side where the technology is flourishing at the same time we have sectional population who are still deprived of basic connectivity. With 5G networks expectations are high not only to provide efficient low power network devices, faster data rates, seamless connectivity and integration of existing technologies through 2G, 3G, 4G, and Wi-Fi but also to provide rural connectivity with zero outage probability so that basic connectivity is assured [7].

The fundamental drivers for 5G networks are:

- Real-time interactions.
- Ultra-high definition videos.
- Critical applications that include medical assistances, traffic management and so on.
- High quality of service, reliability and security.
- Billions of heterogeneous devices.
- Seamless connectivity.
- Diverse services and new evolved use-cases.
- Efficient interoperability between available spectrums.

- Redesigned Air-interface and RAN systems.
- Enormous network deployments with demanding features like Device-to-Device (D2D), dynamic spectrum sharing, self-backhauling and so on.

To fulfill the targets for the 5G wireless networks, the formulated requirements and key components are:

- High Data Transfer Rate both in the Uplink and the downlink
- Improved Spectral Efficiency
- Power Management
- Communication Reliability
- Network Coverage
- Network Deployment
- Network Security
- Tactile internet
- Support for high mobility
- Low Latency ($<$1 ms)
- Energy Efficient Network
- Scalable frequencies to accommodate both low and high data-rate requirements.

2.4.2 Use-cases

With the transition of LTE and LTE-A systems into the future communication system offering plethora of services has defined several new use-cases. These new use-cases being unique in nature will prove to be the corner stones in the 5G networks as they will decide the success and faster adoption of the emerging 5G and how these potential markets can be exploited to have monetary value. Some of the eminent use-cases are defined as [7, 8].

2.4.2.1 Augmented reality

In augmented reality the digital information is blended with user environment in real-time. It exploits wide variety of user experiences and instances. High bandwidth and low latency will be prime necessity for augmented reality.

2.4.2.2 Self-driven cars

As a step toward automated traffic control measures, the vehicles will be enabled with communication capabilities and can sense roads and other co-vehicles to resist accidents. Coordinated vehicles with the traffic control system will enhance the speeds and reduce the risks. High bandwidth, fast data handling and responses in fractions of nanoseconds will be its prime requirement.

2.4.2.3 Video-conferencing and real-time video applications

Real-time video application will emerge as a crucial component in emergency services. It can be used in monitoring applications, security issues, remote medications, identity recognitions and so on. Although 4G systems offer these applications like video conferencing and video monitoring to some extent, 5G networks adds up even more reduced latency and enhanced cloud services.

2.4.2.4 Machine type communication

5G scenario will witness billions of connected devices by the time it is deployed so it becomes critical to have the devices connected and coordinated so as to have seamless services. Evolving 4G offers connected machine type communication subject to enhancements in 5G for more accessibility, simplified usage, flexibility, faster speeds. D2D communication will become eminent specifically in home application systems which include connected devices like smart meters, automated home appliances, smoke detectors and so on.

2.5 How 5G will Change the Society

5G will bring 100 times faster ultra-high speed mobile internet offering speeds more than 10 gigabits per second. This will enables users to download complete movies in 5–6 seconds as per Global Tech and telecom companies report. Apart from speed, 5G will connect billions of new heterogeneous wireless devices ranging from smart gadgets to wearable to embedded industrial products apart from phones, tablets, laptops and many more. The digital offset/Response time popularly known as latency is also guaranteed to be minimised up to 1 millisecond, which is roughly 50–80 milliseconds in current networks. Reduced latency can enable many critical and sensitive applications. Extremely fast response times will make 5G networks to play vital role in traffic automation and control.

Some of the most transformative changes that will be eminent in 5G scenario are namely Rural Connectivity, Universal Internet Access, Unified Air Interfaces and Affordable Broadband.

2.5.1 Rural Connectivity

With the deployment of 5G networks, the world will witness complete transformations. Development will be inevitable as cities will grow into smarter mega cities and so do the villages. Thus having a well-connected communication system will serve as the backbone of the network. Rural connectivity is essentially important.

With the current networks data depicts only 70% of the world's population is under the radar of connected society concept. Rest of the population is still deprived of the basic connectivity. Suggested by a survey report, it is said that by 2020, 85 percent of world's population will be under 3G coverage while 60 percent will be under 4G coverage. 5G networks assure to provide low-cost solutions to provide seamless coverage in rural ends. 5G systems play significant role in making ONE connected world [9, 10].

Rural Connectivity can be achieved via Network slicing efficiently. The network slices can be exploited to provide optimized connectivity depending upon use-cases, applications and users. This technology can efficiently be used for resource management. Another efficient technology to achieve Rural Connectivity will be SDN, NFVs and different virtualization technologies. These technologies provide abstracted services and make network flexible and scalable. Virtualized functions gelled with network slicing is used for network orchestration and enhance network coverage. Carrier aggregation can also be used to provide access to user devices.

The prime hindrance in the path of rural connectivity is the lack of potential economic support per square unit. In rural areas another important hindrance is the uneven population distribution. New hardware will be required to provide rural access.

2.5.1.1 Challenges faced by LTE and other technologies

The indispensable challenge to provide rural connectivity via the LTE services was cost. In absence of required economic standard from the smaller and rural carriers, the larger carriers were dominant in the market. Innovative and combination of access technologies are used to provide broadband and voice services. 4G subsystems incorporated smaller cell concept to meet growing demand for data. This serves as a base layer for developments and deployments of 5G rural concepts.

2.5.1.2 Carrier aggregation for rural connectivity

Carrier aggregation is the technology used to achieve high data rates that increases the transmission bandwidths over single carrier or channel. It is also known as Channel aggregation. It is used in LTE-Advanced networks where it enhanced the transmission bandwidths and bitrates using more than one carrier. LTE-A offers considerably higher data rates than LTE systems. Both Frequency Division Duplexing and Time Division Duplexing support Carrier aggregation for high data throughput. It aggregates, two or more Component Carriers resulting into wider bandwidths. The user end device receives/transmits one or multiple component carriers which may be contiguous and non-contiguous in nature [10].

The access to the user devices can be provided through carrier aggregation of narrow bands. In rural connectivity, mm-waves cannot be used as eNodeB has to provide large coverage thus narrow bands are preferred. Figure 2.3 depicts the rural connectivity using carrier aggregation technique.

In remote areas with least population, the carrier aggregation can be done by a single eNodeB but in areas where more than one eNodeB have coverage the carrier aggregation can be provided by using spatial diversity and combining carriers of all the eNodeB. This approach restricts multiple deployments thus significantly reduces infrastructure costs and maintenance.

In Urban areas, particularly in the indoor environments, user maintained access points provide access in Plug and Play manner. Carrier Aggregation is used to provide sound backhaul to such access points resulting into an energy-efficient network.

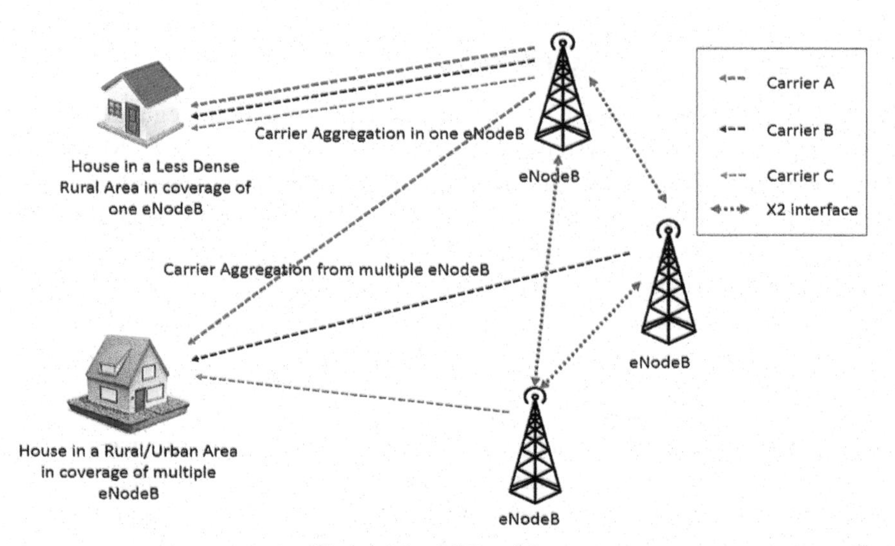

Figure 2.3 5G Rural Areas.

2.5.2 Universal Internet Connectivity and Affordable Broadband

Internet has always been there as an active participant in any technical or non-technical, economic, social etc. growth and development. With the technological uplift of mobile communication serving nearly 70–80 percent of humanity, Internet can be used as "An Enabler to Zero Outage Probability". Universal Internet access can solve coverage problems to nearly 95 percent.

The global standard for 5G networks needs to be formulated at the earliest. The standards include services, co-existing capabilities and most importantly an affordable broadband connectivity of rural communities. The broadband internet connectivity will improve and enhance society both in economic as well as uplift the social and technical standards. Another breakthrough approach to have last-mile connectivity is the Spectrum Sharing, which utilizes white spaces available. The network technologies need to be affordable both in economic as well as availability sense. The spectrum sharing enabled broadband localization and made it available at low costs to support many services like VoIP (voice over IP), video streaming, fast internet access and many more [9, 10].

These emerging technologies are accepted if co-exist with the existing one as an enhancements. The new technologies need to satisfy the mobile ecosystem along with local service providers, existing infrastructures and policies.

2.6 Emerging Technologies in 5G

The 5G networks will form a service defined platform that regulate its availability, connectivity, system robustness, speed and latency. Major requirements for 5G networks are high bandwidth, improved spectral efficiency, minimal latency and seamless integration of existing technologies. 5G amalgamates the evolving technologies independently with the recent technologies being deployed in LTE and LTE-A systems [3, 7, 8]. The brief description of emerging technologies to achieve the defined requirements are:

2.6.1 Massive MIMO

MIMO systems are used to enhance the data transfer rates both at device and network levels. It incorporate multiple antennas at both the transmitter and receiver ends to accommodate more data and ultimately leads to improved performance in terms of reliability, spectral efficiency, and improved radiated energy efficiency. The use of large number of antenna results into high throughput, increased spectral efficiency per unit area, enhanced diversity and compensation for the path loss. Massive MIMO allow high resolution beamforming which are useful at higher frequencies. The antenna elements are designed in a manner to use extremely low power [11].

It also mitigates the intracellular interference using linear precoding and detection methods along with diminishing the effects of noise and fading. Massive MIMO systems can be exploited for MAC layer designing without complicated scheduling algorithms.

The major challenge with Massive MIMO system is the amount of Baseband Processing required for the huge amount of antenna usage and the huge number of antennas demand phase synchronization among them. The Massive MIMO systems suffer from pilot contamination from other cells. Effective Channel Estimation/Feedback required incorporated with fast processing algorithms.

2.6.2 Network Function Virtualization

"Network Function Virtualization (NFV)" is an emerging technology that will set standards for technological revolution in 5G networks. It separates service logic from the hardware platform. The network-based services are enabled by software which formerly requires a specialized hardware [3, 12].

It has brought beneficial transformation both in terms of performance as well as in terms of cost. Its functions are irrespective of location and thus results into a flexible and scalable network with enhanced capacity. It is capable of placing the functions either centralized or near the nodes depending upon the use-cases. It incorporates virtualized services which drastically reduces the investments for the devices. However designing network to efficiently enable services and manage orchestration through software is still a challenge.

2.6.3 Software Defined Network

Software Defined Radio (SDN) is a continuation to the NFV services. It basically separates the DATA and the CONTROL planes. It abstracts functions into virtual services. It configures load and demand through software to maintain the quality of services (QoS) and consumption time [3].

It offers an alternative to the physical infrastructure to manage network in a simple cost-effective approach. It allows different service deployments using the same physical and logical network infrastructure. It facilitates an Open Standard Network where the direct programmability feature separates the data and control planes controlled by SDN controllers. It also let the operators to configure, manage and optimize network resources and regulate traffic load dynamically [8].

However, IT revolution from existing physical structures to software functions has brought various complexities to manage operator side services. With SDNs, its standardization, unified interface and security are major issues. NFV and SDN are already being developed in LTE networks as the fundamental component.

2.6.4 Millimeter-wave

Millimetre-wave (mm-wave) contest as one of the promising solutions to 5G networks. It exploits high frequency band ranging from 20–80 GHz. This technology offers more bandwidths to be allocated to render faster deliveries, high-quality video and to increase the communication capacity [5].

The mm-wave differs from other technologies in terms of high propagation loss, directivity, and sensitivity to blockage. It eliminates the severe path loss by exploiting the high beamforming gain from large number of antenna elements [5, 13].

The challenges associated with the mm-wave technology are mainly interference management, IC and system design, spatial reuse and so on. Despite of challenges it has its remarkable applications in deploying small cells and femto cells and in forming wireless backhaul. It offers both co-located and distributed architectures.

2.6.5 Cognitive Radio

The Cognitive Radio (CR) is a salvation for the scarce spectrum bands. In CR a transceiver intelligently sense the used/unused channels and selects vacant channel on sharing basis for transmission. It acts as the key enabler for self-organizing networks by increasing frequency reuse, spectral efficiency and energy efficiency. The basic challenges associated with the CR are [11, 14]:

- Choice of sensing algorithm: There are a number of sensing algorithms. The Energy detection algorithm, the Matched filter algorithm and so on have their own merits and demerits.
- Concern about latency: 5G aims to provide latency less than 1millisec-onds but in CR sensing time and the time to set-up connection is challenging.

- Handover: The number of handovers in case of CR are predicted to be higher than the licensed spectrum handover and that will increase turnaround time of the network.
- Routing: the routing mechanism still have rooms for development to ensure security.

2.6.6 Heterogeneous Networks

Heterogeneous Networks (HetNet) offers macro, pico or femto cell deployments in an ultra-dense manner to provide maximum network coverage and has its applicability at network level. This reduction in the cell size resulted into increased spectral efficiency and reduced transmission power apart from increased network coverage. The femtocells can be deployed as low power cells in residential and enterprise scenarios whereas the picocells considered as high power cells can be used to provide outdoor coverage and in macrocells to fill coverage gaps. It also facilitate integration of various access technologies from legacy systems such 2G, 3G, 4G, Wi-Fi and so on based on coverage area and proposed topology [3, 4].

Small cells are the corner stones in the HetNet deployment resulting into a flexible and scalable network. The major challenges with the HetNets are Intercell Interference, Distributed Interference coordination due to uncoordinated access points, efficient MAC measures, device discovery, connection establishment and most important is the power management. It can be a potential candidate for 5G networks to ensure zero outage probability in terms of network coverage.

2.6.7 Internet of Things

The Internet of 'everything' (IoT), demand faster connectivity and huge capacity so that it can connect the world with entities. It has changed the concept of connectivity and access. IoT connects people, appliances, vehicles, service based applications and many more with the digital world. IoT forms a major and very important part in the 5G networks as it will seamlessly connects web of users, devices and services [3, 4, 11].

The existing technologies would not be efficient enough to deploy IoT communication and connectivity which dictates higher reliability, vast capacity, faster broadband, connectivity and many more. The applications namely under IoT are Videos services, Rural and urban applications, diverse device connectivity, control and monitoring applications, active response systems and many more.

2.7 Conclusions

5G will the revolutionarily transform the way we are communicating today. It will take things a step ahead of our imaginations. Massive connected devices, high bandwidth, energy efficiency; cost effectiveness and so on will be its characteristic feature. 5G systems will play vital role in forming human centric society where everything and anything will be connected to provide service.

To meet up the various new use cases, evolved requirements and service oriented delivery in 5G things need to be transitioned. The transition from legacy systems including the evolving LTE and LTE-A will co-exist in the new scenario to set an interoperable society.

There are many potential candidates to be exploited to fulfil 5G as a scalable, manageable, and flexible network. The key technologies including multicarrier techniques, modulation, coding techniques and so on have to be blended to give desired outcome. The virtualization will play main role in providing seamless access to billions and trillions of connected devices forming a flexible network.

5G claims to be faster than 4G with greater bandwidths, faster downloads critical applications and so on. Its implementation has to overcome barriers like deployment cost, crowed frequency range, integrated devices and so on. The integrated devices at user ends will be one of the major pull back as it involves evolving existing billions 2G, 3G, 4G handsets to 5G compatible devices. An optimized and adaptive network framework accommodating everything from past to future technologies together with trending needs will mark 5G's success.

5G has to face many challenges before it can be adopted into society. Giants of telecom industries are contending to get 5G standards at the earliest. Existing mobile-broadband technologies will provide sound backbone to overall access technologies. Provisioning Rural Connectivity and Universal Internet Access with affordable broadband will be major challenge before 5G becomes mainstream.

Acknowledgement

Authors would like to thank former researchers from Vishwaniketan namely, Sounak Moulik, Nitish Mital and Sandeep Mukherjee for their contributions that are valuable in preparing this chapter.

References

[1] Prasad, Ramjee. 5G: 2020 and Beyond. River Publishers, 2014.

[2] Wunder, Gerhard, et al. "5G NOW: non-orthogonal, asynchronous waveforms for future mobile applications." IEEE Communications Magazine (Volume: 52, Issue: 2), 2014: 97–105.

[3] Intelligence, G. S. M. A. "Understanding 5G: Perspectives on future technological." London, 2014.

[4] Ramjee Prasad, Global ICT Standardization & 5G Standardization, Volume 1, No. 2, November 2013.

[5] Ernestina Cianca, Mauro De Sanctis, Albena Mihovska, Ramjee Prasad, "CONASENSE: Vision, Motivation and Scope," Journal of Communication and Navigation, Sensiing and Services (CONASENSE), vol. 1, no. 1, January 2014.

[6] Ramjee Prasad, "Human Bond wireless Communications," in Wireless World research forum, Marrakech, Morocco, May 20 2014.

[7] Wen, T., and P. Y. Zhu. "5G: A technology vision." (2013).

[8] ERICSSON WHITE PAPER, 5G SYSTEMS. http://www.ericsson.com/res/docs/whitepapers/what-is-a-5g-system.pdf, Jan. 2015. Accessed: 2015-05-29.

[9] Boccardi, Federico, et al. "Five disruptive technology directions for 5G." Communications Magazine, IEEE 52.2 (2014): 74–80.

[10] Shen, Zukang, et al. "Overview of 3GPP LTE-advanced carrier aggregation for 4G wireless communications." Communications Magazine, IEEE 50.2 (2012): 122–130.

[11] Wang, Cheng-Xiang, et al. "Cellular architecture and key technologies for 5G wireless communication networks." Communications Magazine, IEEE 52.2 (2014): 122–130.

[12] Lu, Kejie, et al. "Network function virtualization: opportunities and challenges [Guest editorial]." Network, IEEE 29.3 (2015): 4–5.

[13] Niu, Yong, et al. "A survey of millimeter wave communications (mmWave) for 5G: opportunities and challenges." Wireless Networks 21.8 (2015): 2657–2676.

[14] Badoi, Cornelia-Ionela, et al. "5G based on cognitive radio." Wireless Personal Communications 57.3 (2011): 441–464.

About the Author

Nidhi is working as research assistant/assistant professor at Vishwaniketan, Navi Mumbai, India since September 2015. She received her Master of Engineering in Electronics & Communication in May 2015 and her Bachelor of Technology in Electronics & Telecommunication in December 2013. She has done different projects and training, namely: Modelling NAN Implementing IEEE 802.15.4g for Smart Grids, Rourkela Steel Plant vocational training for 30 days in Telecom division, Twitter Feed Display using Arduino Due Microcontroller and Hardware Implementation of Fast Multiplier Architectures Using Verilog HDL. She has presented her research outcome in diverse conferences and seminars such as: keynote speech on "5G Security" in 4th International IEEE 5G Summit held at IIT Patna – March 2016, Special presentation on "Security and Privacy" in 4th International IEEE 5G Summit held at IIT Patna – March 2016, Paper presentation entitled "Security Challenges in a 5G Network" in IEEE Global Wireless Summit-2015 and tutorial on TV White space for Affordable Broadband Access and Transmission technologies for 5G in IEEE Global Wireless Summit-2015.

3

Mm-waves Promises and Challenges in Future Wireless Communication: 5G

Maryam Rahimi[1], Hitesh Singh[2] and Ramjee Prasad[1]

[1]Center for TeleInfrastruktur (CTIF), Aalborg University, Denmark
[2]HMR Institute of Technology and Management, Delhi, India

3.1 Introduction to Millimeter-waves

Frequency bands ranging from 300 MHz to 3 GHz are used for radio communication devices, such as TV, satellites, Global Positioning System (GPS), and Bluetooth. To solve the issue of the band getting crowded, researchers have proposed millimetre-waves for next-generation wireless systems. These waves include a large amount of unused spectrum from 30 GHz to 100 GHz and thus satisfy high-quality and high-speed broadband networks required by users and companies. Moreover, mm-waves provide high transmission rate, wide spread spectrum, and immunity to interference because of their large bandwidth. However, the use of mm-waves for wireless communications presents several advantages and challenges. This chapter starts with a brief introduction regarding mm-waves and discusses methods for channel propagation and characterization. The advantages of mm-waves in terms of their bandwidth and capacity are also discussed. The chapter ends with presentation of the application of mm-waves.

Millimeter-waves (mm-waves) assigned to the electromagnetic spectrum correspond to radio band frequencies ranging from 30 GHz to 300 GHz, with wavelengths of 10 mm to 1 mm. These waves are longer than infrared waves or X-rays but shorter than radio waves or microwaves. The high frequency of mm-waves and their propagation characteristics motivate researchers to apply mm-waves for various applications, including transmission of large amounts of data, cellular communication, radar, and so on. An overview on different frequency bands based on united state frequency allocations is shown in Figure 3.1.

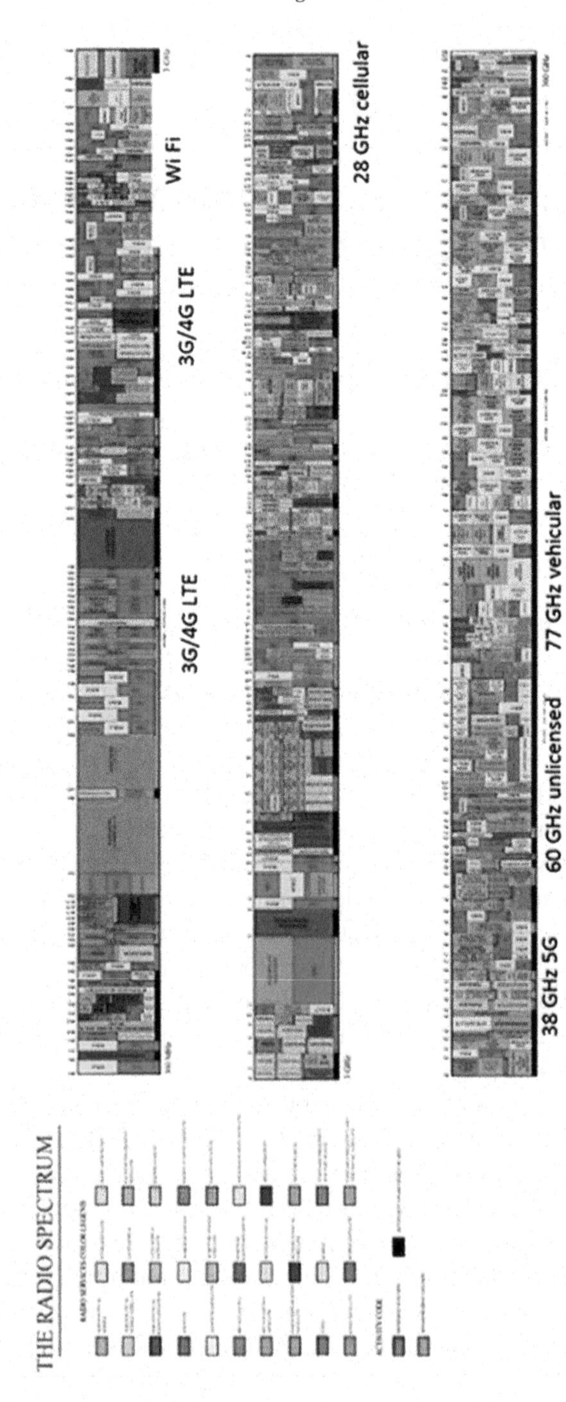

Figure 3.1 United State frequency allocations and mm-waves spectrum.

With its huge amount of available bandwidth, mm-waves are used to transmit large amount of data. Another important advantage of mm-wave propagation is called "beamwidth." This parameter is a measure of the process through which transmitted beam spreads out as it gets farther from its point of origin. Radars take advantage of this specific property of mm-waves. The use of millimeter-length microwaves can help engineers to overcome one of the most important challenges in antenna design. For a given antenna size, beamwidth can be decreased by increasing the frequency and thus, the antenna could also be made smaller.

There are certain advantages in using mm-waves; some of them are listed as below:

- Numerous spectra are available because of less operations occurring at mm-waves.
- Frequency reuse could be done in shorter distance because mm-waves exhibit high attenuation in free space.
- Large antenna arrays for adaptive beamforming can be used for mm-waves.
- Small wavelengths allow reduction in component size, achieve narrow beamwidths, have high resolution, and so on.
- Wide bandwidths, which are around main carrier frequencies (35, 94, 140, and 220 GHz), could provide a high information rate capability; wide-band spread spectrum capability, high immunity to jamming and interference, and so on.
- Extreme high frequencies allow multiple short-distance usages at the same frequency without interfering with one another.

Besides, there are several challenges and open issues about mm-waves, which should be addressed in future studies. For example mm-waves suffer from limited communication range because of atmospheric attenuation (10–20 km), reduced range capability in adverse weather, poor foliage penetration, particularly in dense green foliage, smaller antennas, which collect less energy in a receiving side, thereby reducing the sensitivity and so on. Signals with shorter-wavelength suffer from absorption by fog, dust, and smoke. For example at 60 GHz (5 mm wavelength) oxygen molecules will interact with electromagnetic radiation and absorb the energy. This reaction indicates that 60 GHz is not a suitable frequency for use in long-range radar or communications, because the oxygen absorbs the electromagnetic radiation and signal. Moreover, given that the 60 GHz signal does not travel far before it loses all its energy, this frequency comes in handy for securing short-range communications, such as local wireless area networks used for portable computers, where hackers should not tap into the data stream.

3.2 Channel Propagation of Millimeter-waves

The channel models at mm-wave are different from other frequency bands because the propagation environment has a different effect on smaller wavelength signals. For example, diffraction tends to be lower due to the reduced Fresnel zone. Scattering is higher due to the increased effective roughness of materials, and penetration losses can be much larger. Mm-wave channel models use common properties as low frequency systems (multi-path delay spread, angle spread, and Doppler shift), with different parameters though (few and clustered paths for example leading to more sparsity in the channel). In addition, several new features are introduced to account for high sensitivity to blockages (buildings, human body, or fingers) and strong differences between line-of-sight and non-line-of-sight propagation conditions. Many opportunities use the mathematical properties of sparsity in channel estimation and equalization and recoder/combiner design.

Mm-waves have a special propagation features because of very small wavelength compared to the size of most of the objects and devices in the environment. Understanding these channel characteristics and extracting proper channel model for mm-waves is fundamental to developing wireless network and also signal processing algorithms for mm-wave transmitter and receivers.

In 1998 a measurement on wideband channel at 60 GHz for indoor scenario was carried out to investigate the behaviour of the mm-waves at that certain frequency [1]. The path loss exponent, Ricean K factor and rms delay spread of the mm-waves were extracted from the measurement results. The results show if the transmitter and receiver antenna are aligned and a strong LOS component is present, the K-factor decreases with distance, down to a certain level. Figure 3.2, shows this effect for the measured data presented in [1]. The results also confirm that the rms delay spread (trms) increases with the distance, also to a certain level.

Figure 3.3 demonstrates the changes of the level of the wideband average of the received power with distance at 60 GHz frequency measurement, with omni-directional antenna used. The step was 4 cm. As it is seen, the decay with distance is small. This can be explained with the fact that there are many reflections and that they sum up together [2].

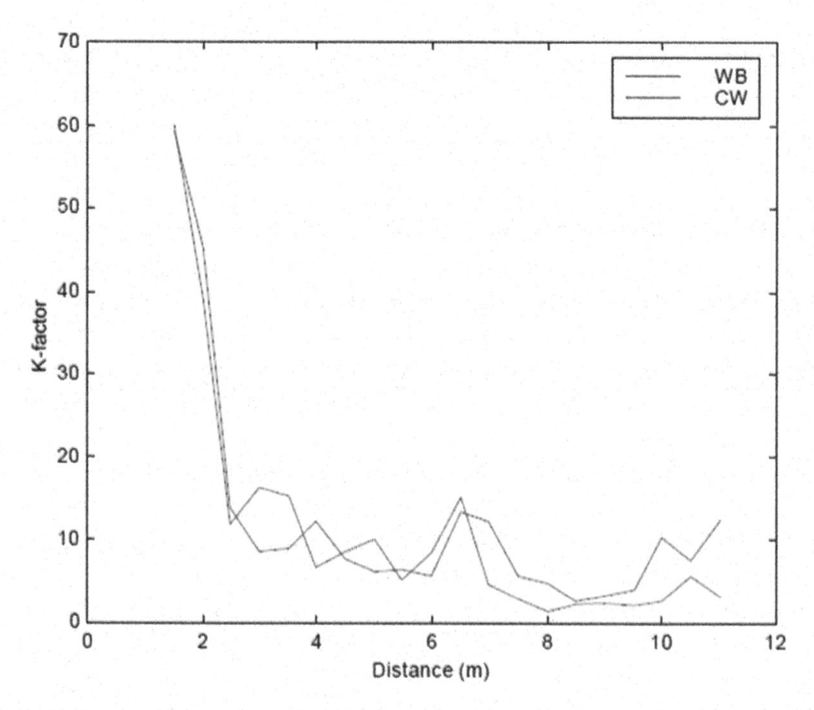

Figure 3.2 The changes of the K-factor in LOS situation in the common room (CW: continuous waves, WB: wide band) [1].

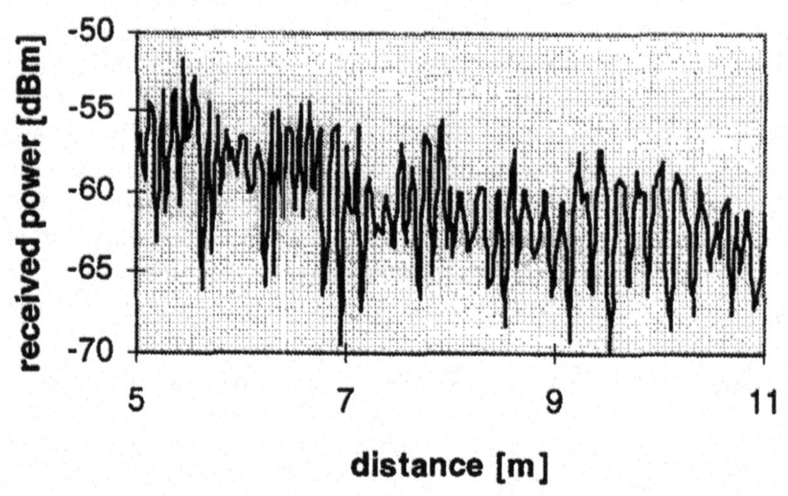

Figure 3.3 Wideband average received power in indoor scenario at 60 GHz [2].

More recent work has focused on path loss models for longer range outdoor links to assess the feasibility of mm-wave pico-cellular networks, including measurements in New York City [3, 4]. An interesting outcome of these studies is that, for distances of up to 200 m from a potential low-power base station or access point (similar to cell radii in current micro- and pico-cellular deployments), the distance-based path loss in mm-wave links is no worse than conventional cellular frequencies after compensating for the additional beamforming gain. These findings suggested the mm-wave bands may be viable for pico-cellular deployments and generated considerable interest in mm-wave cellular systems. At the same time, the results also show that, employing mm-waves frequencies in cellular networks is important for directional transmissions, adaptive beamforming, and other MIMO techniques.

Wideband measurements with 200 MHz of bandwidth discovered that city streets do not cause much multipath, as the rms delay spread was observed to be lower than 20 ns [4]. Measurements and models illustrated that path loss in LOS environments behaves almost the same as free space.

Other outdoor measurements in a city street environment at 55 GHz showed that power decreased much more rapidly with distance through narrower streets compared to a direct path or through wide city streets [5].

Samsung has been active in measuring mm-wave channels for future mobile communications. Initial tests were performed at 28 GHz and 40 GHz to study penetration losses for common obstructions such as wood, water, hands, and leaves [6, 7]. In May of 2013, Samsung Electronics announced the company was able to transmit data up to 1.056 Gbps at 28 GHz over distances up to 2 km by using an adaptive array transceiver with multiple antenna elements [8].

The measurement results are also shown that the range of mm-wave communications is limited because of the rain attenuation and atmospheric and molecular absorption characteristics of mm-wave propagation [9, 10]. Moreover, Oxygen absorption at 60 GHz band has a peak that ranges from 15 to 30 dB/km [11]. The channel characterization in [12] presents that the non-line-of-sight (NLOS) channel suffers from higher attenuation than the line-of-sight (LOS) channel.

Moreover, Electromagnetic waves have weak ability to diffract around obstacles, if the size of the obstacles is significantly larger than the wave-length. With a small wavelength, links in the 60 GHz band are sensitive to blockage by obstacles such as human bodies or furniture. It is shown in the literature; blockage by a human decreases the link budget by 20–30 dB [13].

Collonge et al. [14] carried out propagation measurements in an indoor environment, while human bodies are moving around, and the results show that the channel is blocked for about 1% or 2% of the time for one to five persons. Considering mobility in human bodies plus constant object shows a huge challenge in mm-waves links. Therefore, maintaining a reliable connection for different applications is a big challenge for mm-wave communications.

3.3 Data Rate and Millimeter-waves

The first generation cellular network was introduced and operated in 1978, which was designed for using basic analog systems for voice communications [15]. During the year 1991, 2G was introduced for providing voice and data services with improved spectrum utilization. It was using digital modulation and time division or code division multiple access. During the period of 2001, 3G was introduced with high speed internal access and improved audio and video streaming capabilities. It uses technologies like wideband code division multiple access (W-CDMA) and high speed packet access (HSPA). The 4G of mobile communication was introduced by ITU in 2011 [15]. The technology used in it was the International Mobile Telecommunication – Advanced (IMT-Advanced). Although LTE radio access technology was also used in 4G networks. LTE is an orthogonal frequency division multiplexing (ODFDM) based radio technology which supports up to 20 MHz bandwidth. For enabling high spectrum efficiencies, linked quality improvements and radio pattern adaptation new technology was introduced called Multiple Input Multiple Output (MIMO) [15].

With the tremendous increase of demand for capacity in mobile broadband communications every year, wireless carriers must be prepared for the thousand fold mobile traffic increase in 2020. It forces researchers to find new wireless spectrum which has capabilities to support high data rate demand. The future of mobile communication is 5G technology using mm-waves spectrum [15].

Utilizing mm-waves for 5G wireless technology is a huge step forwards since in the past, mm-wave spectrum was primarily used for satellite communications, long-range point-to-point communications, military applications, and Local Multipoint Distribution Service (LMDS) [16].

The demand for faster and more reliable communication will continue growing at extreme rates, such that annual mobile traffic will exceed 291.8 Exabytes (EB) by 2019 [17]. CISCO has forecasted that mobile data traffic will increase from 2.5 EBs per month in 2014 to 24.3 EBs per month in 2019 [18]. By the year 2020, Nokia and Samsung predict a 10,000x increase in traffic on wireless networks with virtually no latency for content access [8, 19].

The main profit using mm-wave carrier frequencies is the larger spectral channels. For example, at 60 GHz unlicensed mm-wave bands, channels with 2 GHz of bandwidth could be expected. Larger bandwidth channels mean higher data rates, which is the greatest benefit of using mm-waves spectrum in wireless communication.

Besides, massive MIMO is a promising technique for 5G cellular networks. Prior work showed that high throughput can be achieved with a large number of base station antennas through simple signal processing in massive MIMO networks. Massive MIMO promised capability of greatly improving spectral and energy efficiency as well as robustness of the system. In a massive MIMO system, the transmitter and receiver are equipped with a large number of antenna elements (typically tens or even hundreds). Massive MIMO is recommended with mm-waves to overcome challenges of gaining higher data rate. Smaller wavelength captures less energy at antenna due to path loss and so on. Moreover, larger bandwidth means higher noise power and lower Signal to Noise Ratio (SNR). Massive MIMO helps mm-waves to cover those problems and gain higher data rate. Hence, exploiting extra gain from large antenna arrays in the system is promised in massive MIMO.

As a result of the use of large antenna arrays of the transmitter and receiver and combined with radio frequency and mixed signal power constraints, new MIMO communication signal processing techniques are needed. The low complexity transceiver algorithms for wide bandwidths become important. Hence, opportunities abound for exploiting techniques, such as compressed sensing for channel estimation and beamforming.

Precoding and combining is different at mm-wave for three main reasons.

- Parameters have to be configured because of different array. This stage requires different algorithms for finding both analog and digital parameters, and makes the resulting algorithms architecture-dependent.
- The channel is experienced by the receiver through analog precoding and combining. This feature means that the channel and the analog beamforming are intertwined, which makes estimation of the channel directly a challenge.
- More sparsity and structure in the channel result from the use of largely close spaced arrays and large bandwidths. This condition provides structures that could be exploited by signal processing algorithms.

3.4 Application of Millimeter-waves

The applications of mm-waves are enormous in areas such as Wireless Local and Personal Area Networks (WLAN, WPAN) in the unlicensed band, 5G cellular systems, vehicular area and ad hoc networks, as well as wearable devices.

Exploiting frequency bands from 76–81 GHz in different radar ranges for automotive radars has been investigated as one of the major mm-wave applications. WLAN, WPAN, and 5G cellular are being developed with mm-waves. Mm-waves are used in next wireless technology generation to provide high throughput in small cells. There is a huge advantage to utilize mm-waves in MIMO and massive MIMO communication in different applications and scenarios, such as single and multi-user and relay. Vehicle-to-vehicle technology is developing dramatically by using mm-waves, because high data rate for sharing high rate sensors, radar, video, and so on is needed in this technology.

Mm-waves are been used in health science applications for different usages. Wearable devices, such as fitness trackers and smart watches, improve data rate due to usage of mm-waves. The high speed wearable networks provided by mm-waves can connect cell phones, smart watches, augmented reality glasses, and virtual reality headsets [20]. Clearly the future is bright for new applications of mm-waves.

Figure 3.4 shows how the next generation of wireless technology like 5G could evolve human life in terms of healthcare services and devices. Different sensors gather different information from human's body and data loads to the cloud and been monitored by care givers. The solution and help from medical center could be achieved with no delay. Thanks to mm-waves for their wide band to make this dream to the reality.

Mm-waves have plenty of potential applications as well. For example, with the recent interest and research towards communication between cars and data centers also autonomous vehicles, mm-waves may play a role in providing high data rate connections between cars. This feature is natural because mm-waves already form the backbone of automotive radar, which has been widely deployed and developed over the past 10 years [21]. The combination of mm-wave communication and radar is another interesting application [22]. Mm-waves could enable high rate and low latency connections to clouds that allow remote driving of vehicles through new mm-wave vehicle-to-infrastructure applications.

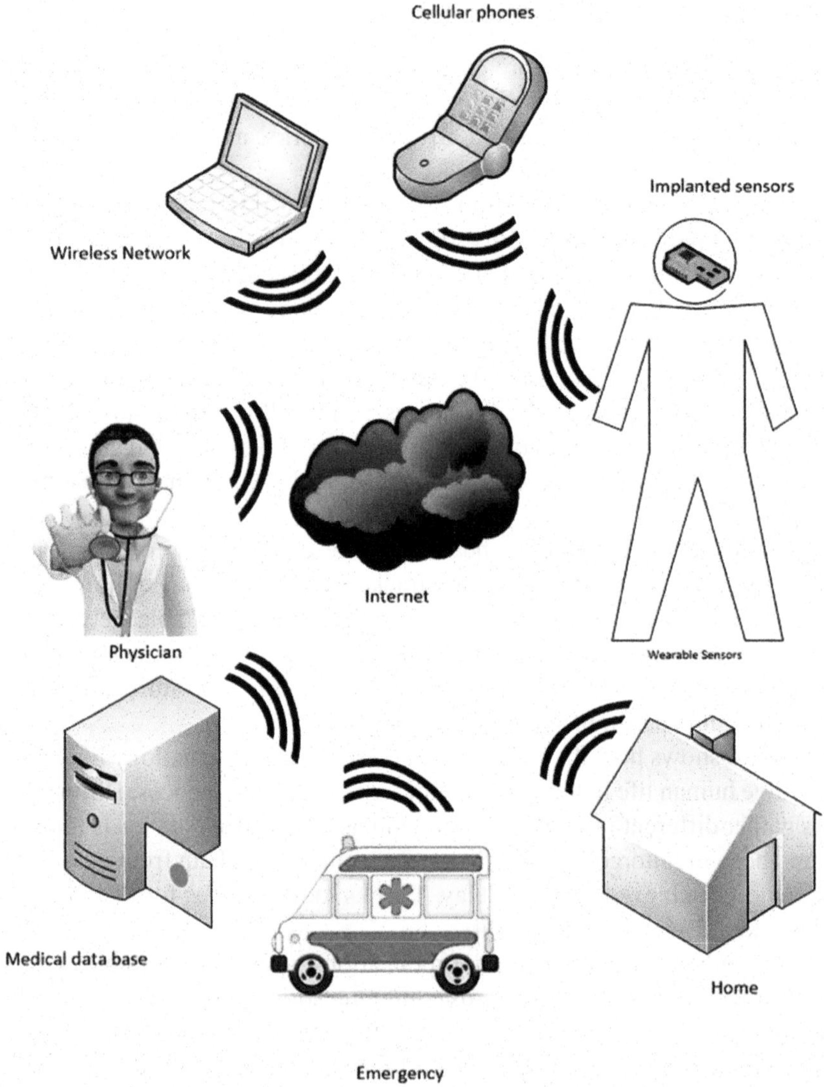

Cellular phones
Implanted sensors
Wireless Network
Internet
Physician
Wearable Sensors
Medical data base
Home
Emergency

Figure 3.4 Future of medical services using mm-waves in 5G wireless communication systems.

Figure 3.5 shows an overview on how cars are going to communicate to each other and to the base stations in future. It is clear the network infrastructure should be developed for fast and reliable communication, which is needed in the roads. That is one of the 5G promises with help of mm-waves.

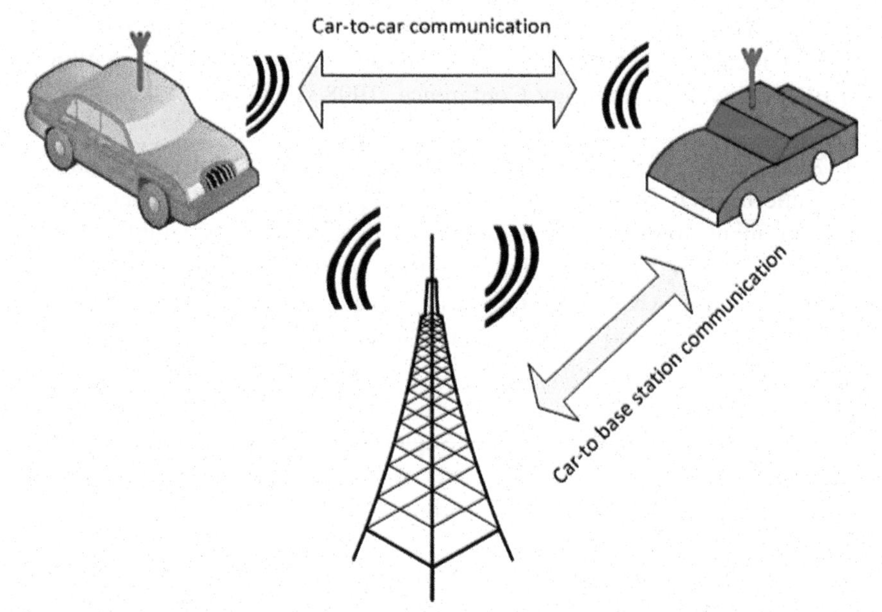

Figure 3.5 Vehicle-to-vehicle communication, and vehicle-to base station communication promised by 5G and mm-waves.

3.5 Conclusions

Mm-waves are as old as wireless technology. The first experiment was conducted by Bose in 1895. Since then, use of mm-waves has been limited for many years. However, the rapid development in wireless communication technology has resulted in giving mm-waves more potential to be part of the future of wireless communication specially 5G technology. The greater capacity besides unique characteristics of mm-waves increases the applications of mm-waves. However, many challenges and open issues should be addressed by researchers concerning the new physical technology, software-defined architecture, measurement of network, and so on to promote the development of mm-waves in wireless communication.

References

[1] J. Purwaha, A. Mank, D. Matic, K. Witrusal, and R. Prasad, "Wide-Band Channel Measurements at 60 GHz in Indoor Environments," in proc. IEEE benelux 6th Symp. Veh. Technol. Commun, Brussels Belgium, vol. Oct. 1998.

[2] D. M. Matic, H. Harada, and R. Prasad, "Indoor and outdoor frequency measurements for Mm-waves in the range of 60 GHz," in 48th IEEE Vehicular Technology Conference, 1998. VTC 98, 1998, vol. 1, pp. 567–571.

[3] Y. Azar, G. N. Wong, K. Wang, R. Mayzus, J. K. Schulz, H. Zhao, F. Gutierrez, D. Hwang, and T. S. Rappaport, "28 GHz propagation measurements for outdoor cellular communications using steerable beam antennas in New York city," in 2013 IEEE International Conference on Communications (ICC), 2013, pp. 5143–5147.

[4] P. F. M. Smulders and L. M. Correia, "Characterisation of propagation in 60 GHz radio channels," Electron. Commun. Eng. J., vol. 9, no. 2, pp. 73–80, Apr. 1997.

[5] H. J. Thomas, R. S. Cole, and G. L. Siqueira, "An experimental study of the propagation of 55 GHz millimeter waves in an urban mobile radio environment," IEEE Trans. Veh. Technol., vol. 43, no. 1, pp. 140–146, Feb. 1994.

[6] S. Rajagopal, S. Abu-Surra, and M. Malmirchegini, "Channel Feasibility for Outdoor Non-Line-of-Sight mm-wave Mobile Communication," in 2012 IEEE Vehicular Technology Conference (VTC Fall), 2012, pp. 1–6.

[7] H. Zhao, R. Mayzus, S. Sun, M. Samimi, J. K. Schulz, Y. Azar, K. Wang, G. N. Wong, F. Gutierrez, and T. S. Rappaport, "28 GHz millimeter-wave cellular communication measurements for reflection and penetration loss in and around buildings in New York city," in 2013 IEEE International Conference on Communications (ICC), 2013, pp. 5163–5167.

[8] T. S. Rappaport, W. Roh, and K. Cheun, "Mobile's millimeter-wave makeover," IEEE Spectr., vol. 51, no. 9, pp. 34–58, Sep. 2014.

[9] Z. Qingling and J. Li, "Rain Attenuation in Millimeter-wave Ranges," in 2006 7th International Symposium on Antennas, Propagation EM Theory, 2006, pp. 1–4.

[10] R. J. Humpleman and P. A. Watson, "Investigation of attenuation by rainfall at 60 GHz," Proc. Inst. Electr. Eng., vol. 125, no. 2, pp. 85–91, Feb. 1978.

[11] R. C. Daniels and R. W. H. Jr, "60 GHz wireless communications: emerging requirements and design recommendations," IEEE Veh. Technol. Mag., vol. 2, no. 3, pp. 41–50, Sep. 2007.

[12] S. Geng, J. Kivinen, X. Zhao, and P. Vainikainen, "Millimeter-wave Propagation Channel Characterization for Short-Range Wireless Communications," IEEE Trans. Veh. Technol., vol. 58, no. 1, pp. 3–13, Jan. 2009.

[13] S. Singh, F. Ziliotto, U. Madhow, E. Belding, and M. Rodwell, "Blockage and directivity in 60 GHz wireless personal area networks: from cross-layer model to multihop MAC design," IEEE J. Sel. Areas Commun., vol. 27, no. 8, pp. 1400–1413, Oct. 2009.

[14] S. Collonge, G. Zaharia, and G. E. Zein, "Influence of the human activity on wide-band characteristics of the 60 GHz indoor radio channel," IEEE Trans. Wirel. Commun., vol. 3, no. 6, pp. 2396–2406, Nov. 2004.

[15] R. Prasad, 5G:2020 and Beyond, vol. 2014. River Publisher.

[16] S. Y. Seidel and H. W. Arnold, "Propagation measurements at 28 GHz to investigate the performance of local multipoint distribution service (LMDS)," in IEEE Global Telecommunications Conference, 1995. GLOBECOM'95, 1995, vol. 1, pp. 754–757.

[17] T. S. Rappaport, "Keynote Speech: Millimeter-wave Wireless Communications – The Renaissance of Computing and Communications," presented at the IEEE International Conference on Communications (ICC), Sydney, Australia, vol. Jun. 2014.

[18] "Cisco visual networking index: Mobile data traffic forecast update, 2013–2018," presented at the CISCO, San Jose, CA, USA, vol. Feb. 2014.

[19] "FutureWorks NSN White Paper: Looking ahead to 5G," presented at the Nokia Solutions and Networks, The Netherlands, vol. Dec. 2013.

[20] Pyattaev, K. Johnsson, S. Andreev, and Y. Koucheryavy, "Communication challenges in high-density deployments of wearable wireless devices," IEEE Wirel. Commun., vol. 22, no. 1, pp. 12–18, Feb. 2015.

[21] H. H. Meinel, "Automotive radar: From its origins to future directions," Microw. J., vol. 2013, no. 9, pp. 24–28.

[22] C. Sturm and W. Wiesbeck, "Waveform Design and Signal Processing Aspects for Fusion of Wireless Communications and Radar Sensing," Proc. IEEE, vol. 99, no. 7, pp. 1236–1259, Jul. 2011.

About the Author

Hitesh Singh is presently working as Assistant Professor in HMR Institute of Technology and Management, Delhi affiliated to Guru Gobind Singh Indraprastha University, Delhi. He has done his Master in computer Science at Indraprastha university, Delhi. Hitesh was involved in several projects and research and he has more than 12 International and National publications. He is currently pursuing his Ph.D. at the Technical University of Sofia, Bulgaria under GISFI program.

4

The Fog over the Meadow and the Cloud in the Blue Sky

Rajarshi Sanyal

Belgacom International Carrier Services, Belgium

4.1 Introduction

It is said that 'Change Is Tough But Constant Evolution Is Invigorating'. Last few years, we have witnessed various nuances of telecom network architecture during its continuous evolution trail. The network moved from pure premise based deployments towards the implementation of virtualised instances on the cloud. Technologies like SDN and NFV were introduced which fuelled this network metamorphosis. The evolution to a virtualised network framework had benefited the network operators by reducing the capital expenditure, streamlining the operational processes, reducing the time to deploy new functionalities and rapidly scale up the network capacity. Typically in such scenarios, the network operations are spawned in the cloud environment, the physical location of the clouds are apparently not important anymore. This is indeed one of the selling points of the cloud based technologies. However from a purist's standpoint, this may not be spot-on. The Cloud that we are talking of is not certainly thin mist. The network functions require physical processing resources to operate and the cloud has to fuel its needs. These virtual entities will have to be hosted in real physical machines located in some data centers. The distance between the user devices and the virtual machines have an impact on the data latency and the spatial dimension of the backhaul loop, which determines the overall quality rendered for the given service. The regulatory requirements of certain countries can restrict the user data traffic flow within its boundaries. This makes it overtly important for us to comprehend where the virtual environment is actually hosted, within the country or outside its peripheries. In spite of these constrains, a cloud based mobile network meant for human users appears to be tenable. In fact, it has many 'pros' compared to the COTS (Commercial Off-The-Shelve) installations. However for the machine to machine devices, those simple choices cannot be made. Some machines may need brisker data processing. Some machine may invoke large volumes of queries which can contribute to an avalanche at the backbone network. The centralised network environment in the cloud may be less responsive to such massive computing requests. It is imperative that a part of the processing functions needs to be brought closer to these machines. The telecom researchers thus proposed to lug the cloud from the sky to the ground to mitigate some of these concerns. The cloud morphed into the fog, thus the name 'fog networks'.

The Fog network concept was coined by Cisco [1] and is well within the scope of the 5G network paradigm. Recently a new industry group, The OpenFog Consortium, was formed to define and promote fog computing. The consortium, founded by ARM, Cisco, Dell, Intel, Microsoft and Princeton University in November 2015, seeks to create an architecture and approach to fog, edge, and distributed computing.

FogNets connect a plethora of devices like wearable devices, connected vehicles, sensors, actuators etc. It provides access to internet via aggregate router and also empowers the devices to communicate amongst themselves in a collaborative fashion. These devices form many "local clouds" at the edge of the network. Contrary to the notion of cloud computing, the fog computing renders capabilities at edge entities with an endeavour to localise computing and relieve the cloud core from the prosaic tasks of repetitive data processing related to control signalling, monitoring, optimisation or even social networking. The edge can also have storage capabilities which is utilised to store chunks of data which can be consolidated to compile the big data. The delta tasks that cannot be accomplished by edge entities are handled by the cloud.

This chapter is organized in 5 sections including introduction (4.1). Section 4.2 provides some background on Fog network with some practical examples. Section 4.3 discusses the overall architecture consolidating fog and the cloud entities Section 4.4 sets out the attributes of the fog. Section 5.5 concludes.

4.2 Background and Examples

Imagine a scenario where a group of college students are travelling together. We see them bantering, sharing pictures and other media files amongst themselves, as well with friends outside their group via social networking platform. The physical environment hosting the social media application may be thousands of kilometres away. The data from the individual smartphones traverses thousands of kilometres every time to reach the central application server which determines the identity and location of the target device. If the originator and the target are in vicinity as in our case, then the trajectory of the data packets becomes a big loop. The traffic trombones via the central server and arrives back at the target device which may be just a meter away. When we have such millions of users exchanging peer to peer information, the backbone traffic and the computing resources expended at the core can be substantial. The pertinent question here is, why cannot the devices communicate with each other at the edges without involving the core? We cannot do it today because most of the device clients are not designed presently to take up a share of the processing tasks done at the cloud core. They are also not fabricated to actuate collaborative communication amongst them. The core applications at present are also not modelled to dispel some processing tasks to the edges, implying the mobile devices.

With the advancement of semiconductor technologies, the edge devices are becoming more intelligent. The network ecosystem is slowly adapting to empower the edge devices to take up some processing load to alleviate the burden of the core, and to make the application more intuitive, real time and responsive. Here are some practical illustrations.

4.2.1 Uber Fog Network

As per the traditional approach the Uber smartphone application of the Uber driver communicates with the Uber backend to continuously update the trip information. The primary server collates the data and replicates across the hot backup servers in other data centres. In case the primary server breaks, the secondary server takes over. As the primary server continuously communicates with the Uber mobile devices and simultaneously needs to update the secondary servers, we envisage substantial data communication between all these elements. Apart from the backhaul and front haul load, we also need to understand that the computing power required in the core to choreograph these processes is substantial. The system is also subject to latency and we need to keep a watchful eye there to avoid timeouts of data transactions.

To circumvent this situation and optimise the ecosystem, Uber has come up with a unique proposition, i.e., to empower the devices to trigger the computing at the edge and diminish the processing burden of the core network (Figure 4.1). The device application has been redesigned. It consumes additional computing resources and memory of the smartphones of the cab drivers. The trip data is locally stored in the device itself which takes decisions unilaterally without invoking any triggers in real time to the core network. The device is also used as a storage device where moving vector information and network mutation information can be stored temporarily. Uber could eventually demonstrate that even if the whole core network is out of service, or isolated from the cluster of end devices, the network operations can still run as normal and unperturbed, thanks to the collaborative edge computing realised by the mobile device processor and its memory. We must acknowledge that this practical demonstration is a giant step towards establishing the fact that the fog concept can indeed work.

4.2.2 IFTTT and Google OnHub

Another novel approach of fog computing is the IFTTT (If This Then That) technology which is presently supported by Google. With this technology, users can orchestrate processes within the mobile device across various applications residing on the handset to trigger a user defined automation task. Before IFTTT was born, this was done at the core network by a mobile service delivery platform which used to house a process choreographer to actuate network driven automation tasks achieving the alike results. With IFTTT, this automation process has been pushed towards the edge. Not only can IFTTT work within a device, but can also help to create a fog network at the edge via an intelligent router with some computing capabilities. Once such example is

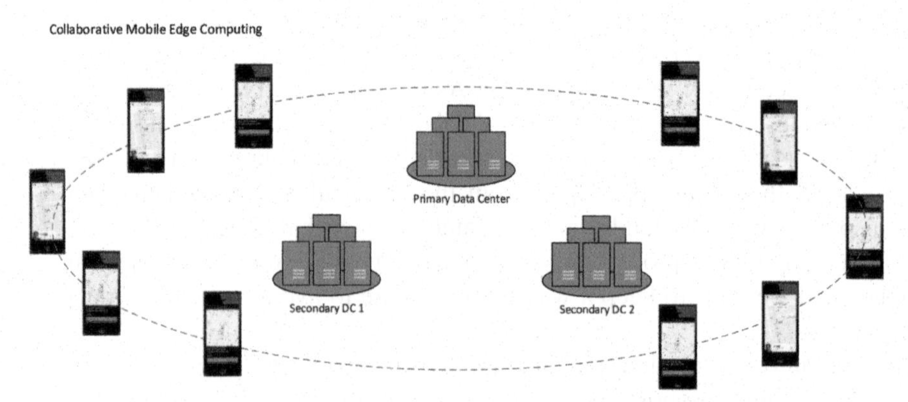

Figure 4.1 Collaborative computing at the edge in Uber network.

Google's Onhub which supports IFTTT and help in connecting a plethora of devices at the edge like smartphones, Internet of Things.

IFTTT can be triggered when devices connect and disconnect from OnHub (Figure 4.2). The user can program any custom built logic, termed as recipe, over a smart interface in their device. It lets users manage and prioritize Wi-Fi to connected devices through an app. Some examples are provided by Onhub on what we can do with it:

- Build a recipe that turns your WiFi Bulb wirelessly. The lightbulb disconnects automatically when you leave your room, actually when your smartphone disconnects from WiFi network.

Figure 4.2 High level representation of the Google OnHub.

- Trigger IFTTT to send you an intimation by email or SMS when your child gets home from school and his/her phone connects to the home's OnHub network.
- Automatically prioritize Wi-Fi to your Chromecast over other existing wifi device connections, when the Chromecast connects to your OnHub network after you plug it in.

4.2.3 Smartgrid

The Power Smart Grid is not a new concept. It aims to pilot the control information flow between the power plants more efficiently, optimizes power distributions, diagonizes and isolates circuit failures, scale up, scale down power level dynamically, actuate smart power distribution etc. The 'As is' power grid is modelled to fetch information from the edge actuators and sensors towards the central cloud to pivot the processing tasks. With the surge in the power requirements, the smart grid network architecture is becoming more intricate. Issues like data privacy, scalability, high availability and latency are more potent than ever. The existing smartgrid model needs to adapt to fit in the new environment. The smartgrid designers are looking towards the fog. In South America, SkyWave IDP satellite terminals control and monitor Smart Grid applications. The SkyWave M2M devices have inbuilt capability to actuate analytics at the edges. The thresholds and filters are implemented at these edge devices so that they can preprocess and filter out the relevant information to be sent to the cloud core for group processing. Hence only the critical messages, like changes in current, voltage and power factor information are conveyed to the core network.

4.2.4 Edge Analytics

An American petroleum company had to abstract essential information like valve pressure, volume of extracted gas, general health of the machines, etc. Earlier, the actuators were disseminating information towards the central application server in the cloud, where the big data was post-processed subsequently for analytics. Latter on, the analytical capabilities were added at the edge, and only vital data like instantaneous high and low pressure and flow, incremental/decremental gas volume, out of range values were being sent wirelessly to the cloud. The company could increase the production by 30 per cent.

Apart from the above examples there are many other uses cases. Few which comes to my mind are smart cities, rail safety systems connected cars, connected wearables, smart traffic lights, sensors for military applications, aircraft sensor network, Virtual Radio Access Networks etc.

4.3 Fog Network Architecture and Its Attributes

Fog network comprises of dense computational virtual structures that provide computing, storage, and networking services at the network's edge. Some of the characteristics of such platforms include low latency, location awareness and operability in wireless hetnet access environment. We shall provide more details on the attributes of the Fog network later in this chapter.

Cisco's impression of Fog Network architecture is in Figure 4.3.

As evident from Figure 4.3, the IoTs and M2M devices will be attached to the fog nodes. The application which governs the fog data services can be run on a separate platform at the edge or can as well run within the fog nodes in a distributed fashion. The data is analysed and sieved by the fog application, and only essential data is transferred to the cloud. The temporary data is retained at the fog layer else which can be used for daily operational work without communicating with the cloud.

The latency drop off is the key to Fog Networks. Figure 4.4 provides a landscape of the latency requirements of Business Intelligence and analytics pertaining to the fog and the cloud network in purview of a typical operation business, say where data is collected from actuators of a petroleum factory.

The fog network directly gathers data from the sensors and the actuators and executes application specific analytical data in real time. The typical time to process the M2M (only) data is in milliseconds. As explained earlier, fog network has memory and retains the generic data. Only critical data is conveyed towards the cloud network for centralised processing and book keeping.

As we go up in the analytics chain (as in the diagram), we observe that the time to process the analytical data increases exponentially. It is also interesting to see that the more we move towards the cloud zone, the use cases are no more only confined to M2M, but it find its use in HMI (Human to Machine Interface).

4.3.1 Fog Network in the Context of 5G

The scope for 5G network does not confine itself in just providing superfast connectivity to users. The scope extrapolates to broader domains. Some of them (though not exhaustive) are as follows:

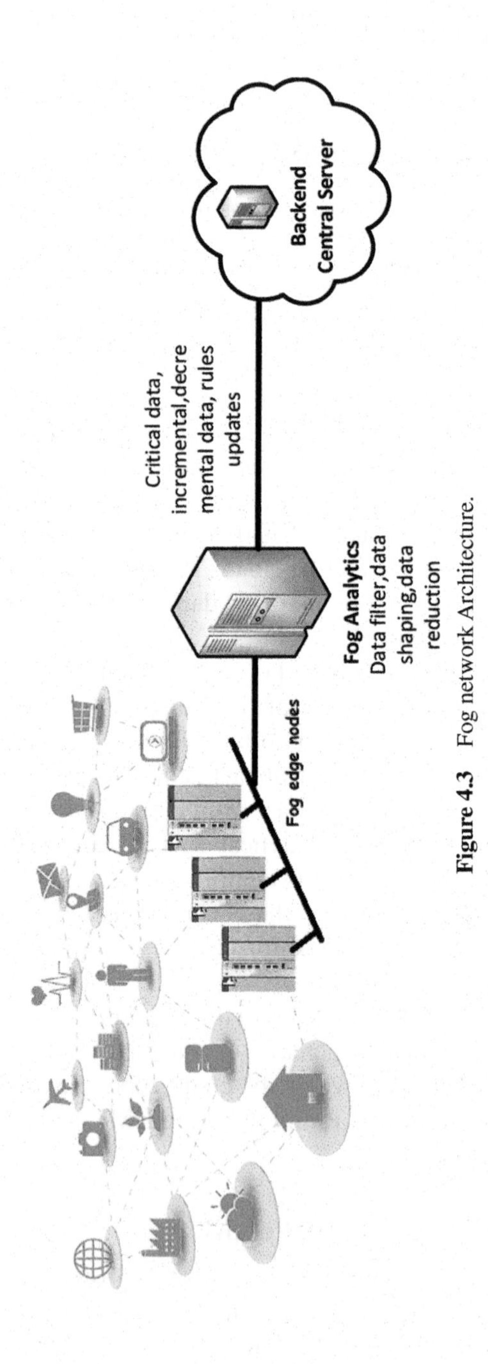

Figure 4.3 Fog network Architecture.

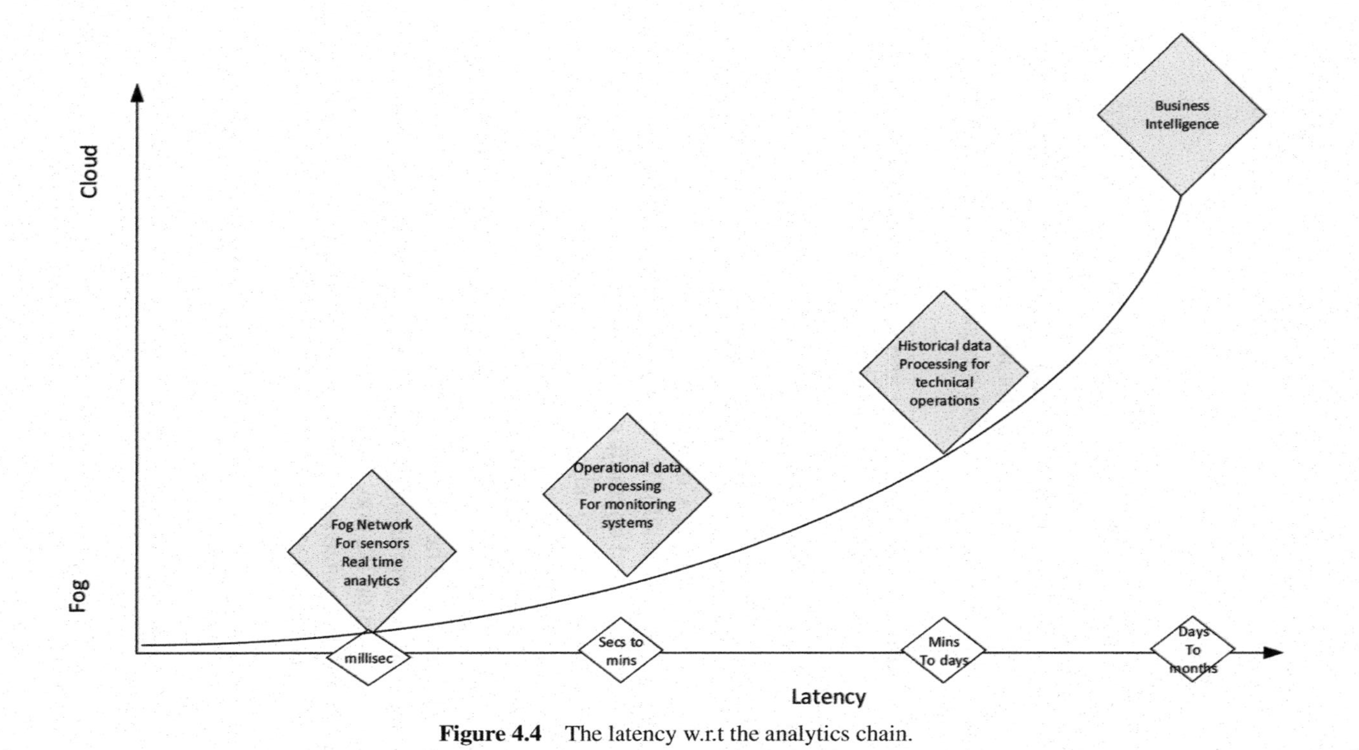

Figure 4.4 The latency w.r.t the analytics chain.

- Integration of the service logic and containers in the network that processes the big data.
- Aggregation of the data feed from the actuators that are linked to varied industrial implementations.
- Cater traffic from human users.
- Comply with the 5G latency requirements.

5G vision for ultra-low latency as envisioned by Ericsson is "To support such latency-critical applications, 5G should allow for an application end-to-end latency of 1 ms or less". Nokia 5G research group has spoken about "Zero latency gigabit experience" for its 5G networks.
Fog network will facilitate this by proposing edge computing in three network domains.

- **Fog Computing**: Assign computing and storage resources at the edge, filter and parse critical data to cloud to reduce backhaul traffic and processing.
- **Fog Mesh Network**: Utilize a swarm of Collaborative device clients at the edges to perform some operations that presently happen in cloud.
- **Fog Radio Access Network (F-RAN)**: Realise real-time Collaborative Radio Signal Processing, Collaborative Radio Resource Management at the edge devices, flexibility to scale up and scale down the access network capacity based on traffic conditions, cognitive radio capabilities that renders the ability to adjust the radio parameters based on traffic conditions and radio environment, and diminish fronthaul payload and processing requirements at the LTE Core.

The fog network and the cloud network landscape in purview of 5G paradigm is captured in Figure 4.5. The 5G LTE-A mobile network core in conjunction with the virtual RAN (comprising of virtual BS pool and RRUs) shapes up the 5 network [2]. The virtual BS pool conveys are user traffic to the core network via backhaul links. The voice and messaging network service are rendered by the IMS core hosted on the cloud platform. This network is designed to serve human users.

Mobile network was initially not designed to serve the machines. Following the evolution trail from 2G to 5G, we witness that the basic network processes, for example location management, addressing, session setup, almost remain the same. The Fog network is typically designed to cater M2M devices. The network is physically isolated from the mobile core network just because the network characteristics are very different due to the machine type usage. The network design and modelling challenges are different here. The traffic patterns we see in some M2M scenario (non-exhaustive) are as follows.

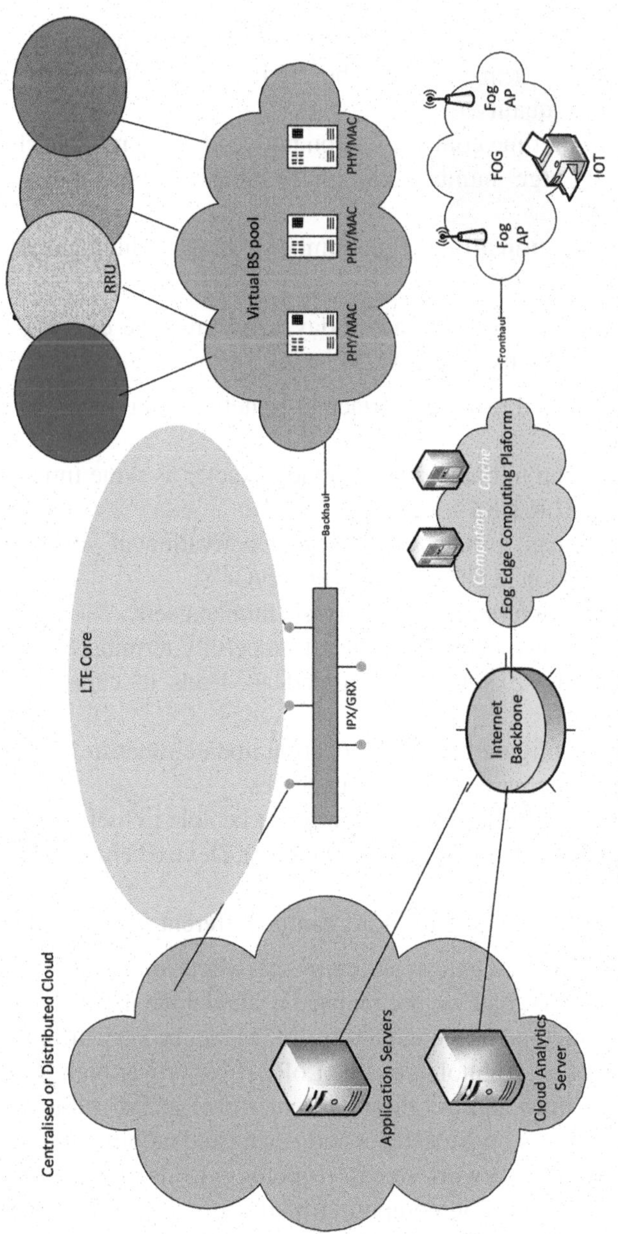

Figure 4.5 Fog and cloud in 5G network paradigm.

- **Electric meters**: Small amount of data, few times a year, no mobility, large number of devices.
- **Health**: Small amount of data, limited mobility, high quality.
- **Medical imaging**: Limited usage but high data volume, needs reliable data channel for quality.
- **SOS**: Once in lifetime communication but needs very reliable channel.
- **ATMs/POS**: Large numbers, small volumes, bursty traffic secure channel.
- **Vehicular communication**: High number of devices, high mobility, low but constant data rate.

Some of the challenges that we face today in sharing the existing mobile network for both human users and M2M are as follows.

1. Lot of devices need to remain attached to network, but remains mostly in idle mode.
2. Many devices can initiate network attach exactly as same time creating congestion at HLR/HSS.
3. Many devices can initiate GTP context activation at same instance creating congestion at GGSN/PGW/PCRF/OCS.
4. Network congestion due to M2M affects human users.
5. Hung GTP sessions: GTP sessions not gracefully terminated by M2Ms while new GTP sessions are initiated → leads to capacity loss at GGSN/PGW.
6. Not properly configured periodic location update timer in M2M device which leads to increase in Location updates.
7. Current device addressing scheme may not be able to fuel growth.
8. Support of BYOD (Bring Your Own Device): Device behaviour can vary widely.
9. Security threats pertaining to M2M can be different from human users.

Any surge in M2M traffic can overwhelm the whole mobile network and impact the human users, had we not realised a standalone network for M2M traffic. The access method implemented for M2M can be very different from a 5G cellular networks. Technologies like LoRa (Low power high Range) [2] have been brought in to address the requirements of M2M traffic. Another new technology based on physical layer addressing and mobility management SMNAT (Smart Mobile Network Access Topology) [3] is ideal for the M2M type implementation. The Fog computing platform is connected to the Fog Edge network serving the edge actuators through the fronthaul pipe. LTE-M (LTE for machines) is also a good candidate for M2M access.

The man and machine also need to communicate and interoperate. A central network orchestrator can play the role to choreograph the processes across the two kinds of environments.

4.3.2 Fog Network Attributes

The key attributes of Fog network are as follows:

Cognizance of Location: The Fog network brings in location awareness, and holds the logic for processing the meta data at the edges in accordance to the location of the diminutive cloud. For example, in gaming sort of applications, the latency requirements achieved depends on the physical location of the fog network which is in vicinity of the gaming equipment.

Heterogeneous operational environment: The fog devices can come in various shapes, sizes and behaviour. There is not yet a common framework that binds them together. So interoperability, federation and backward compatibility are some of the areas where the fog researchers have focus. The few companies which are in the fog consortium are attempting to bridge these differences.

Hetergeneous Wireless access: The device can acquire any kind of network, be it cellular, LoRa, WiFi, Zigbee etc. The fog network should harmonise the traffic before feeding the data to the fog computing platform by the fronthaul channel.

Device Mobility: The devices can be dynamic in nature, cross countries or continents. The technologies like LoRa support mobility in a limited way. To attempt ubiquitous mobility, more futuristic technology like SMNAT can be adopted.

Real Time Analytics: The Fog Edge network should be able to perform analytics in real time so as to impart only pertinent information to the cloud in an attempt to diminish the backhaul load and processing weight at the core network.

4.4 Summary and Conclusions

Cloud computing and fog computing play a complementary role in propelling the evolution of the network ecosystem catering the needs of smartphones and M2Ms. The norms widely vary across the two worlds, i.e., man and the machine, and so is their traffic pattern and data distribution. So while we aim to achieve a degree of seclusion between these two network kinds, it is also imperative that interoperability needs to be attained between them. The Fog edge network initiates local data analysis and picks up critical data bearing the changes to the rules of the game. This delta information is transferred towards the central cloud to reduce traffic and processing there. New generation of interfaces are coming up, new architectural choices needs to be addressed. This includes billing framework, security models, operational and analytics processes, session management etc. It is time to think more about our smart watches and the connected coffee machine rather than confining our thought processes only to cloud, NFV, SDN and virtualization. Fog network bears the potential to bring closer the three worlds, the man, the machines and the cloud network.

References

[1] Fog Computing and the Internet of Things: Extend the Cloud to Where the Things Are: CISCO Whitepaper, 2015.

[2] Shao-Chou Hung; Hsiang Hsu; Shao-Yu Lien; Kwang-Cheng Chen, Architecture Harmonization Between Cloud Radio Access Networks and Fog Networks, Access IEEE, 2015.

[3] A technical overview of LoRa® and LoRaWAN™, Technical Marketing Workgroup 1.0, LoRa Alliance.

[4] Sanyal Rajarshi, Prasad Ramjee, Enabling Cellular Device to Device Data Exchange on WISDOM 5G by Actuating Cooperative Communication Based on SMNAT, Journal of International Journal of Interdisciplinary Telecommunications and Networking, Volume 6 Issue 3, July 2014 Pages 37–59, IGI Publishing Hershey.

About the Author

Rajarshi Sanyal is telecommunication engineer with 19 years of industry experience in the field of mobile communication. He started his career as a telecom engineer in Hutchison in the mobile switching domain in India and later on worked on R&D projects on SS7 and CTI applications. He was involved in R&D in a French Telecom startup in the role of a Domain Specialist – Telecom focusing on the development of new solutions related to CAMEL, SMSCs, Mobile Service Nodes and Mobile Payment solutions etc. He was associated with Reliance Communications as Engineering Manager on Mobile Core Networks and later on as Managing Consultant at IBM responsible for designing the Service Delivery Platform Solutions for Mobile Networks. He is presently working as a Network Architect (Voice and Mobile Data core network engineering) at BICS (Belgacom International Carrier Services) at Brussels, Belgium. BICS is one of the largest wholesale voice, signaling and data carriers worldwide. For the last 8 years at BICS, he had been spearheading tier one projects based on roaming hubs, GTP solutions, LTE instant roaming, Voice over LTE, WebRTC, M2M and IoT technologies.

He holds several EU, US and Indian patents in the field of Mobile and Computer communications. He has published 24 papers in international telecom journals and conference proceedings in US, Europe and Far East. He holds a Ph.D. from Aalborg University, Denmark. In his thesis he proposed a new generation multiple access technology based on intelligent physical layer to conceive a smart mobile network and realize device to device communication within the framework of 5G mobile network architecture.

5

Adding a New Dimension to Customer Experience, the Reality of 6th Senses – 5G and Beyond

Jeevarathinam Ravikumar[1] and Ramjee Prasad[2]

[1]Stanley Advanced Technology Solutions (SATS), UK
[2]Center for TeleInfrastruktur (CTIF), Aalborg University, Denmark

5.1 Introduction

The evolution of 2G to 3G and now the current 4G technologies have changed the user's expectations. The demand for providing the best user experience is ever increasing. The following terms; Customer Experience (CX) and User Experience (UX) are used interchangeably throughout this chapter.

5G and beyond are at an increased pressure of enriching the customer experience beyond what the current generation of technologies can offer. CX will continue to remain an important focus in defining and ultimately deploying 5G and beyond networks and associated solutions.

Consumer devices of today can offer the best in class voice, data and video communications connecting anyone from anywhere at any time. However, the devices of tomorrow will offer virtually all the features as much as a wireless network can offer that means users will have more autonomy in defining what they want and how they want it.

Looking into the evolution of UX amongst various technologies, it is clearly evident that the focus of wireless networking has been concentrated on the end user experience.

The UX evolution can be represented in a very simplified fashion,

$$\sum_{QoS}^{CX} \int f(Network)$$

The function of the wireless network will be fully focused on providing the best user experience. Hence, any evolution in the wireless network era should be built based on the way customer experience requirements will evolve from current to future needs. Evolution is not about quality of service rather it is all about customer experience.

Figure 5.1 shows how the experience has been changing along the wireless network evolution. It has been a paradigm shift moving from 1G to 5G wireless network evolution.

First generation mobile networks (1G), were a real revolution in providing wireless voice from anywhere but with a limited mobility. 2G with GPRS, and i-mode initiated another revolution in mobile communication providing data over wireless with a true mobility. SMS and roaming are also the major user experience revolution in 2G. Subsequently smartphones, especially iOS made a major revolution in user experience by not only introducing a gesture based UI but also introduced the App Store which vastly expanding the functionality of the device. Android and the Google Play Store have helped to penetrate mass markets and phones with larger screens have enriched multimedia applications. Users can enjoy a HD quality entertainment seamlessly while being mobile with LTE-A offering 300 Mbps speed [3].

As the device screen improves not only in size but also in terms of resolution, such as 4K video quality, creating a compelling need for high resolution projection from the customer device.

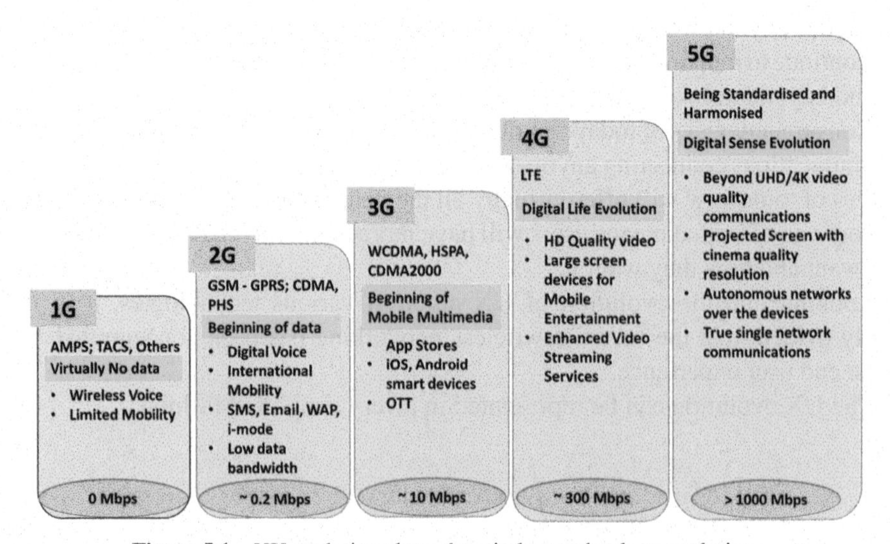

Figure 5.1 UX evolution along the wireless technology evolution.

5G and beyond will offer enormous bandwidth beyond 1 Gbps which can potentially create a demand for devices featuring high resolution video projection equivalent to a 100" virtual screen experience to become a commercial reality. Figure 5.1 shows how users had different experiences during development steps of the mobile communication systems.

5.2 Does the Bandwidth Matter?

Advances in processing power, computing engines, machine learning, nanotechnologies and beyond, and increasing variety of applications in diverse fields will contribute to a demand for increased bandwidth. Eventually, bandwidth evolution in mobile communications will progress towards terabit mobile communications. Today's 4G – LTE user devices offer 300 Mbps speed on a LTE-A compliance handsets.

However, for seamless continuity of HD/UHD video streaming over the air, the network needs to adapt to provide a cinematic standard of streaming that can be beamed over a larger screen or wall or to provide a virtual reality experience. Video bandwidth requirements will be ever increasing as the number of associated devices for each individual increases and as everyone and everything becomes more closely connected. Cinematic quality of video will have a major application in medicine [6].

Referring to Figure 5.2, associated things will demand more bandwidth as most of the associated devices will evolve from low bit rate sensors of today to high bit rate sensors of tomorrow. For example, if the contents of a home fridge are to be seen in real-time for the purpose of updating food stocks, a high resolution video content needs to be transmitted. Increasingly associated things will bring about a major change in human life and way of living, way of working and will also make a major impact on the way business is conducted.

It is now becoming a necessity to be able to deploy a single but fully converged network pipe to match the bandwidth demands and provide seamless wide area coverage. This will eliminate the need for multiple telecom networks and will evolve into a single 5G and beyond network that will ultimately connect everything together from end to end. This will ultimately open up the single wireless conduit from the end users or end nodes and it will eliminate the need for any disparate last mile wired/wireless networks. For example, the video content of the fridge can be directly transmitted over the wide area network of 5G and beyond rather going through a WiFi home network or through a wired Ethernet.

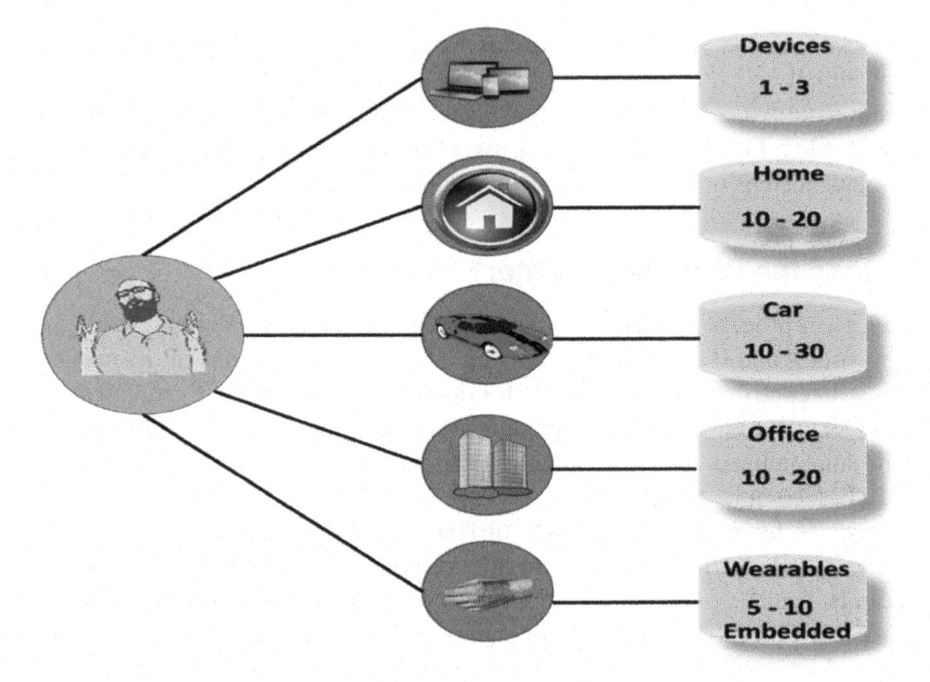

Figure 5.2 Associated things.

- Person with chain of association
- Associated things will be touching close to a trillion in the next decade

5.3 CX of Today

CX of today is still predominantly focused on the network behavior experience rather than the experience at the end user device or the node. Obviously smartphones have more powerful processing capabilities and features that enhance the customer experience of today. Social networking has changed the traditions of human networking to a great extent.

Referring Figure 5.3, the devices of today still do not have much autonomy in offering network functions as a standalone device.

CX of today is focused more on:

- Quality of Experience
- Quality of Service
- Bandwidth

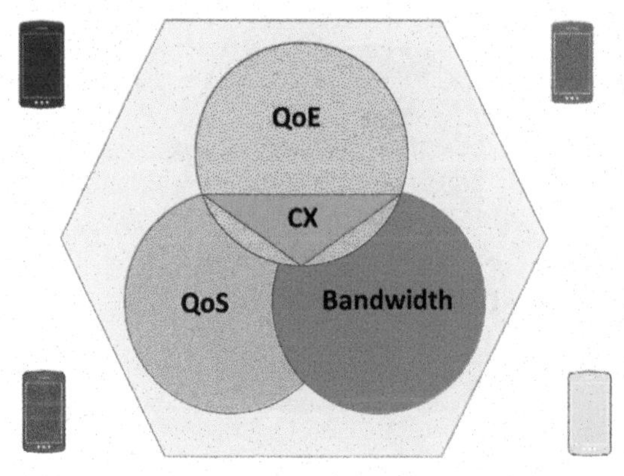

Figure 5.3 CX of today.

5.4 CX of Tomorrow

CX of today has been aimed at delivering experiences limited to a human's five senses, but CX of tomorrow shall go beyond the five senses.

What can be expected beyond the five senses?

A digital sense termed the sixth sense (6th Sense) that will not only predict and prescribe but also provide

- On-demand wisdom provisioning
- Rapid analysis of a situation
- Privacy and Personalization

6th sense and Digital sense will be used interchangeably throughout this chapter.

Referring to Figure 5.4, the network of tomorrow will pass on some network functionality to the device. Devices will have network functions in a way that a smart device can form an autonomous network around the associated things and multiple devices. Hence a complete chain of networks can be formed to provide seamless communication even during an outage due to natural disasters or similar emergency conditions. Essentially, it will be a type of UserCell which provides most of the base station functionality.

CX of tomorrow shall be focused on

- Distributed network capability
- Autonomous CX
- Advanced features with augmented reality (AR) and more

There is going to be an enormous potential for CX and it will grow beyond imagination as the CX will be the only continuing killer application over the

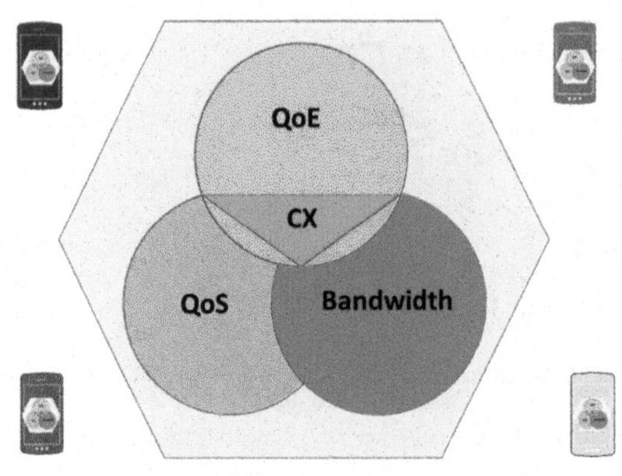

Figure 5.4 CX for tomorrow.

communication networks. Reliable Gigabits bandwidth will be required for a real time high resolution AR platform.

5.5 CX Applications

This section discusses a few of the CX applications that will open up a new era in customer experience. Availability of larger bandwidth on demand, speed, display, and multimedia technologies shall make the user devices much more powerful on top of the autonomous network capabilities built in.

Devices of today offer seamless communication over voice, video and data with the mushrooming of many messaging applications such as iMessage, Whatsapp, Viber, Skype etc. Unfortunately, the quality of the video and the experience of being together still feels incomplete over the channels of today.

Networking needs to go beyond the simpler social networking of today to a greater extent of being "Globally together", regardless of whether it is networking between friends, family or corporations. Hence there is a need for a seamless, simple and instantaneous Globally together ENhanced with Digital Sense (GlobENDS). GlobENDS will virtually end the borders of this world and shall be a lution in human communication.

5.5.1 Virtual World – Home without a Border

Imagine a cooking lecture conducted at America with virtual class rooms in Africa and India. Figure 5.5 depicts the scenario of HD quality live networking directly from a user device. A person sitting in Australia projecting all the video feeds directly from his phone to the three sides of the wall in his room. This is live networking using a 5G and beyond smart device with optical beam projection features. This requires very high resolution video as well as seamless cross-switching.

The same concept can be applied to family networking. Now imagine a scenario involving an extended family with parents, sisters, brothers and families living far apart.

With the advancement in the digital eye wear, a virtual screen of over 100" will make the projection better suited for personal communications.

With GlobENDS, users should be able to have their breakfast together and discuss family matters over the dinner table whilst being in their own place across continents.

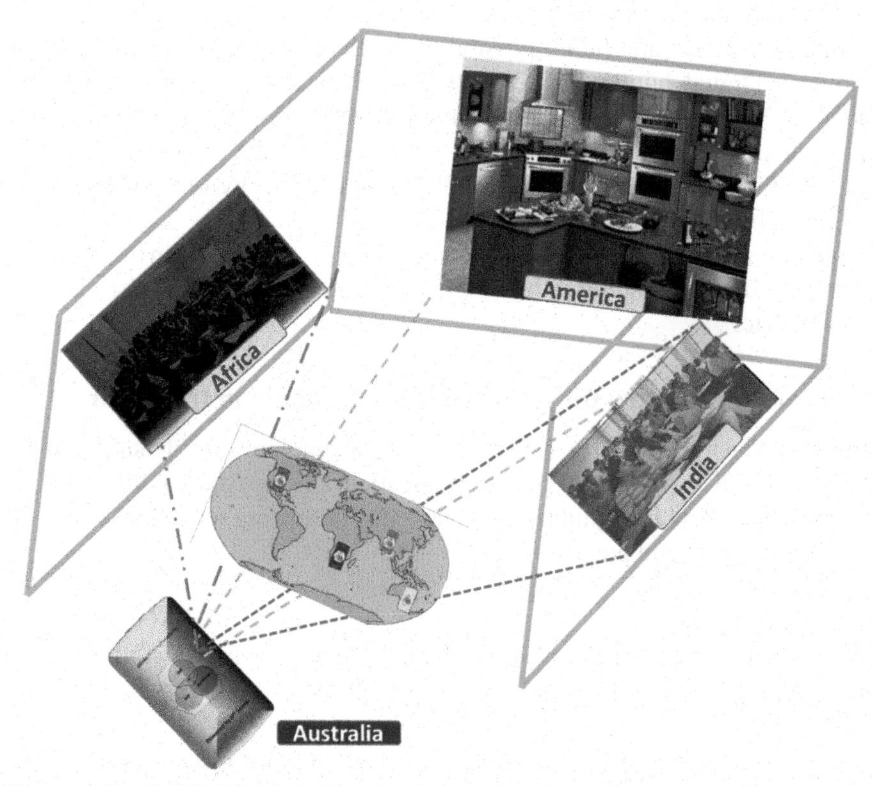

Figure 5.5 GlobENDS, Live HD networking across the continents with a Smart device projecting over three sides of the walls.

5.5.2 On-demand Digital Sense Provisioning

A Major dream for any human being is to obtain wisdom provisioned over the air. Wisdom provisioning shall be a major breakthrough in human evolution.

Wearable and implantable devices are becoming increasingly commercialized with the advent of Nano technologies, biomedical technologies and advanced medical research. With the evolution of advanced implantable devices and neurological research, it shall be feasible for wireless provisioning of wisdom in the near future.

6th sense of wireless provisioning shall open up all the possibilities of innovations in this century. By nature, humans are 'analog', with eyes, ears and mouths that are all processing analog signals which are digitally processed in the human brain.

Research over the possibility of reading the human brain is approaching reality with advancement in neuro science [4].

Figure 5.6 depicts the future possibility for the evolution of the Analog human to a Digital human. WiSense that shall basically enable wireless wisdom provisioning (WWP) through an on demand basis over wireless to the implanted device of the human. This will open up the convergence of Ear & Mouth with digital Senses & Reflections enabled by 6th sense, the digital sense [1, 5].

Application of WiSense shall open up a major revolution in treating intellectually disabled and other brain related conditions such as Alzheimer, Parkinson's etc.

5.5.3 Evolution of How things shall Communicate in the Future

Machine to Machine is not a new concept or technology but it has been around for decades. What makes communication amongst things different with the advancement in digital technologies? The following sections discuss and analyze the past and present.

5.5.3.1 Descriptive

When Alexander Graham Bell invented telephony, it was the first revolution in connecting people through machines but was over the wire – Interconnected over Wires (IoW).

When Marconi invented wireless communications, it was again the first revolution in connecting people through the machines but over wireless – Interconnected over Wireless (IoW).

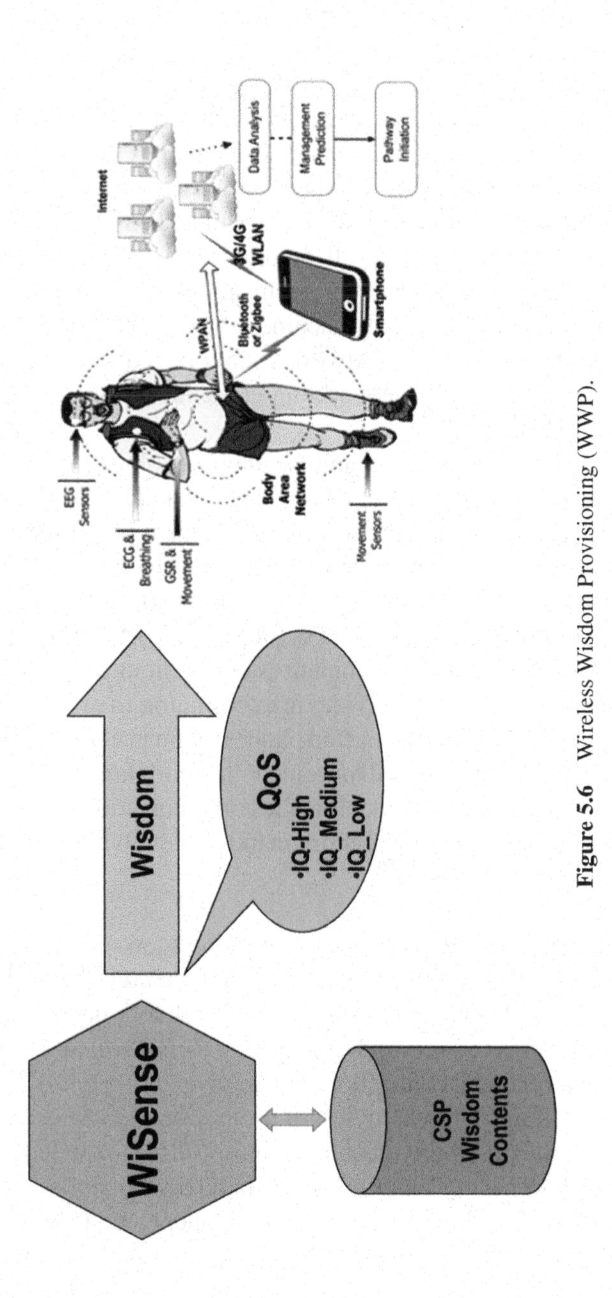

Figure 5.6 Wireless Wisdom Provisioning (WWP).

When NASA launched the first communications satellite, it was again the first revolution in making the possibility of connecting people globally – Interconnected over Satellite (IoS).

When ARPANET was created and subsequently adopted TCP/IP, it was again the first revolution in connecting networks – Internet of Networks (IoN).

When Mobile telephony was made and subsequently Cellular communications became popular, it was again a first revolution of communication from anywhere and anytime – Interconnected over Cellular (IoC).

Then came 1G, 2G and beyond, all predominantly voice communications transmitted through packet data communications which existed not for connected people but for connected machines. SMS and Global roaming are the highlights of GSM/CDMA networks. 2.5G paved the way for data communications over the cellular networks and Blackberry email became the first real killer application.

But Apple's iOS and App store created the major revolution in connecting people, the easiest way of using data over the cellular. This is a real IoS – Internet of Smartphones. Then came Android that made it possible to smartly connect most of the population of the world in a cost effective way. The App store and the Play store have become the killer applications, those who did not make it to the stores went out of business (of phones). Evidently Over The Top (OTT) services have started to dominate as connections are saturated but their attributes such as Speed and Services are continuing to evolve.

Today we have over 6 billion connected phones. Human communication is evolving towards high bandwidth, advancement and enrichment of customer experience. Connecting things and enabling the communication of things with humans shall open up enormous applications in the ever evolving digital space.

5.5.3.2 Predictive

The global population is not going to double anytime soon which means that there is a finite number of possible connections. Hence connections need to go beyond the people. IoT is connecting anything whether living or nonliving. If we associate 100 things to every individual per day and organize them into subsets of things, the potential connections are going to be beyond 600 Billion and possibly past trillions. Ultimately it will hit the Smart Dust dream. This opens up almost infinite possibility of connections. OTIoT – Applications and Services On Top of Internet of Things (OTIoT) shall play a major role and will change business models and the way business is conducted. This is nothing but the OTT of today but in a more powerful and multi-fold method.

5.5.3.3 Further enhancing CX with cross communication with things

Internet of human and things (IoHT) shall open up many applications. Figure 5.7 shows an application in which connected shoes can remind a person, "Don't wear me today since you need shoes with spikes as it is going to snow after two hours", or it might even become smart enough to be provisioned to be a spiky shoe for that period of the day!

Referring Figure 5.7, a dog can be digitally enabled to pre alert that there is a jam at exit A and so it may be better to take exit B as they will be communicating live! Practically anything and everything will start communicating and the level of smartness shall be provisioned by the OTIoT services.

Figure 5.7 Applications – Internet of Human and Things (IoHT).

Aakash Vaani, announcement from the heaven in ancient Indian mythology can be made a reality. A prediction can be made to such an extent that a person can be alerted so precisely, that if he or she takes a particular route at a particular time, he or she shall meet with an accident. This is because of all the things that are communicating and time synchronized with the traffic pattern & behaviour, making a scientific prediction of astrology a reality!

5.5.3.4 Prescriptive

GOD (Generator; Operator; Destroyer) made the living things. Man made the non-living thing, the machines such as aeroplanes etc.

Getting on to making the non-living things communicating among themselves and also with the living things, Man is becoming GOD of devices.

Hypothetically, man can make the corpses, the non-living things communicable and also can preserve the bodies with the wearable suit. With the historical data that have been collected over the living period can be made available to the wearable suit and can make the non-living body completely emulating the living body. It shall be nothing but a Human Disaster Recovery (HDR).

Additionally, extra knowledge can be provisioned to make it smarter than it was. In conclusions, a new sense is being created which is the Digital Sense that will eventually become the 6th sense of the human being. Although the 6th sense will be manmade, let us try to put this to good use to help make this world a happy connected, safe and more enjoyable place for all to enjoy. Digital innovation through connecting things should be embraced by everyone so as to stay connected with the growth of business in the new economy.

5.6 Conclusions

Digital Sense, the 6th Sense shall be the future of the digital evolution. 5G and beyond shall make the 6th sense a reality, eventually leading to major advancements in the human experience and create major applications in healthcare to home automation.

References

[1] N. Ravikumar, N. H. Metcalfe, J. Ravikumar, R. Prasad, "Smartphone Applications for Providing Ubiquitous Healthcare Over Cloud with the Advent of Embeddable Implants", Wireless Personal Communications, Volume 86 Issue 3, February 2016.

[2] R. Prasad, "Human Bond Communications (HBC)". Wireless Personal Communications, 87(3), 619–627, 2016.

[3] J. Ravikumar, R. Prasad, "Challenges in guaranteeing Lifestyle Enrichment while evolving towards 5G", 18th GISFI standardisation series meeting, New Delhi, 2014.

[4] Vikash Gilja, Paul Nuyujukian, Cindy A. Chestek, John P. Cunningham, Byron M. Yu, Joline M. Fan, Mark M. Churchland, Matthew T. Kaufman, Jonathan C. Kao, Stephen I. Ryu, Krishna V. Shenoy, "A high-performance neural prosthesis enabled by control algorithm design", nature neuroscience 15, 1752–1757 (2012).

[5] J. Ravikumar, R. Prasad, "Wireless 2020", CTIF, Aalborg University, 2006.

[6] E. C. Prakash, L. Qiang, A. Anwar, W. K. Chia, N. Ravikumar, and T. K. Chiong, "A Practical Surgery Simulator", Proceedings of the International Conference on Information and Automation, December 15–18, 2005, Colombo, Sri Lanka.

About the Author

Jeevarathinam Ravikumar is Technology Director (Europe) – Stanley Advanced Technology Solutions (SATS) with Stanley Security Europe of Stanley Black & Decker group.

He designed India's first wireless telephony for rural India and was involved in designing Digital Radio/Switches working for Centre for Development of Telematics, Government of India. He was one of the pioneers in starting WCDMA research in Singapore way back in 1998 while working for ASTAR, Singapore. He has been advising wireless telecom operators on 3G/4G deployment and mentoring technology startups. He has over 25 years of experience in wireless communications with the blend of technology and business expertise and worked for many multinational technology companies and also founded his own startups.

He has Masters in Communications and Computer networking from Nanyang Technological University, Singapore and Bachelors of Engineering from Annamalai University, India. He is a senior member of IEEE. He holds international patents in wireless coverage solution and published papers in Beyond 3G & 4G networks/services.

6

IMT for 2020 and Beyond

François Rancy

Radiocommunication Bureau, International Telecommunication Union, Switzerland

6.1 Introduction

Work on International Mobile Telecommunications (IMT) has been ongoing for nearly three decades in ITU. This has been an open process which has included ITU's Member States, national and regional standards development organizations, equipment manufacturers, network operators, as well as academia and industry forums.

With the development of the IMT-2000 3G mobile systems and the current deployments of the IMT-Advanced 4G systems, this activity has revolutionized the way people communicate around the world. ITU is now working together with these partners in the same open process to develop the standards for IMT-2020.

A vital element in the standardization process has been the identification and global harmonization of frequency bands for the operation of IMT, thereby enabling interoperability, roaming and global economies of scale.

This Chapter outlines the process and timeline for standardizing IMT-2020, and the current status of this work. It also provides a perspective of the key technical requirements and spectrum aspects to address to finalize the standard in 2020.

6.2 Background

The second generation of mobile telephone systems were developed in the late 1980's and initially deployed in the early 1990's. Certainly the transition from the first to the second generation of mobile phones was characterized by the change from analog to digital communications, but it was also characterized by the growing requirement for these sytems to operate on a regional, if not global, basis.

Regional/global operation of these systems was hampered by having multiple incompatible standards as well as different frequency bands and channel arrangements being used in different parts of the world. This in turn had a significant impact on the cost, and hence affordability, of these systems. Recognizing this, the ITU membership established a group of experts to study the requirements of future public land mobile telecommunications systems (FPLMTS).

Studies on FPLMTS were conducted in the CCIR (the former ITU-R) Interim Working Party 8/13, with the first substantive output being a decision by the 1992 World Administrative Radiocommunication Conference to identify specific frequency bands for the operation of FPLMTS. The studies then focussed on developing the set of detailed radio interface specifications for FPLMTS.

ITU-R Task Group 8/1 was established to develop these 3G radio interface specifications, which were finally approved in May 2000 in Recommendation ITU-R M.1457 – "Detailed specifications of the terrestrial radio interfaces of International Mobile Telecommunications-2000 (IMT-2000)" [1]. The name change from FPLMTS to IMT and the principles and process for the further development of IMT were established by the ITU Radiocommunication Assembly 2000 in ITU-R Resolutions 56 [2] and 57 [3].

ITU-R Working Party 5D was subsequently established to continue the work on IMT. In close collaboration with the relevant national and regional standards development organizations, a yearly update process for IMT-2000 was applied to cater for the evolution and enhancement of the standard. ITU-R Recommendations were also developed to address the implementation aspects of IMT-2000 such as global circulation of terminal equipment, radio frequency channel arrangements and sharing studies between IMT and other radio services. In parallel, successive World Radiocommunication Conferences have periodically identified additional frequency bands for IMT to cater for the rapidly growing demand for mobile communications, particularly mobile broadband data.

At the same time, Working Party 5D initiated work to address the need for a global platform on which to build the next generations of mobile services – fast data access, unified messaging and broadband multimedia: IMT-Advanced. The IMT-Advanced radio interface specifications were finalized in 2012 and are specified in Recommendation ITU-R M.2012 – "Detailed specifications of the terrestrial radio interfaces of International Mobile Telecommunications Advanced (IMT-Advanced)" [4]. These 4G systems are currently being deployed throughout the world, and it is expected that these systems will continue to evolve and be enhanced in the coming years.

With an eye to the longer term requirements, in 2012 Working Party 5D then commenced studies on the next phase of development: IMT-2020. It is planned to finalize the IMT-2020 specifications in the year 2020.

6.3 IMT-2020 Standardization Process

The process for developing the IMT-2020 radio interface specifications is shown in Figure 6.1 and is similar to the process that was successfully applied in the development of the IMT-2000 and IMT-Advanced standards. It is important to stress that the development of the IMT standards is not carried out by ITU alone. It is a highly colloborative process with substantial input from and coordination with all involved national, regional and international standards development organizations, partnerships and fora.

The first step of the process was to establish the vision for IMT for 2020 and beyond, by describing potential user and application trends, growth in traffic, technological trends and spectrum implications, and by providing guidelines

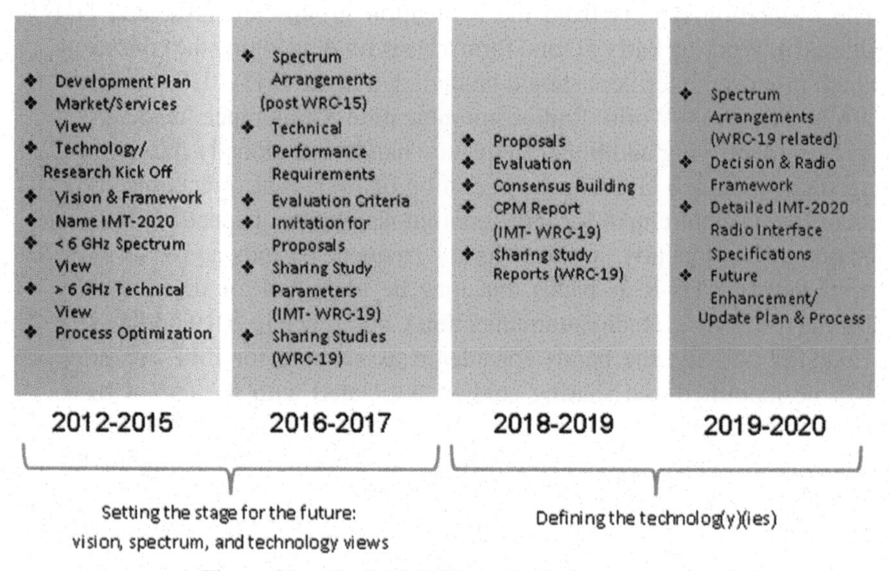

Figure 6.1 The IMT-2020 standardization process.

on the framework and the capabilities for IMT for 2020 and beyond. This work reached a significant milestone with the approval of Recommendation ITU-R M.2083 – "IMT Vision – Framework and overall objectives of the future development of IMT for 2020 and beyond" [5] in September 2015.

The ITU has now issued the invitation for submission of proposals for candidate radio interface technologies for the terrestrial components of the radio interface(s) for IMT-2020 and invitation to participate in their subsequent evaluation in Circular Letter 5/LCCE/59 [6] on 22 March 2016.

The next step, planned for 2016–2017, has now commenced, with studies being undertaken to review how best to make use of the spectrum identified for IMT at world radiocommunication conferences (WRCs), including the recent WRC-15 held in November 2015. It will also be necessary to establish the detailed technical requirements and a set of criteria to support the evaluation of proposals for the radio interfaces for IMT-2020.

The submission of proposals is expected to begin in October 2017 and end by mid-2019. The evaluations against the criteria will then be carried out by independent evaluation groups established for this purpose, and participation in these groups is not limited to ITU members. In the past there has been a very good level of participation in these evaluation activities with the active involvement of Administrations, equipment manufacturers, network operators and academia.

The evaluation reports from the evaluation groups are presented and considered in Working Party 5D and form a basis for developing the consensus on which proposed interfaces should be included in the IMT-2020 standard.

While the 2015 World Radiocommunication Conference made good progress in identifying additional frequency bands and globally harmonized arrangements below 6 GHz for the operation of IMT, it also recognized the potential future requirement for large contiguous blocks of spectrum for this application. Consequently, it called for 11 frequency bands above 24 GHz to be studied by ITU-R as bands that may be identified for future use by IMT at the next World Radiocommunication Conference in 2019 (WRC-19). As a parallel activity, the bands considered as suitable for IMT operation need to be identified and sharing studies associated with the use of these bands need to be conducted in preparation for WRC-19, and these decisions regarding spectrum use will need to be taken into account in developing the final IMT-2020 specifications.

The overall timeline for the standardization of IMT-2020 is presented in Figure 6.2, with final approval of the IMT-2020 specifications expected in around the 3rd quarter of 2020.

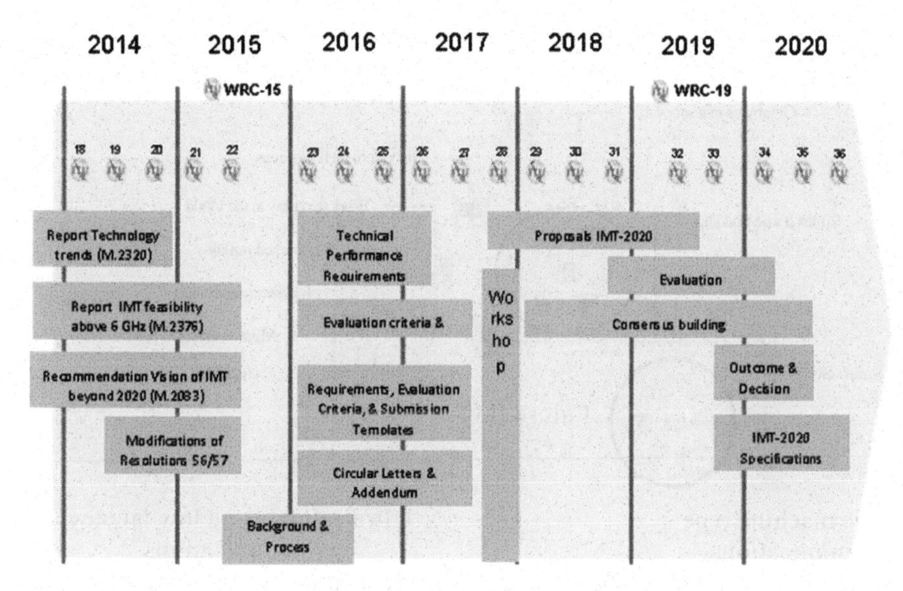

Figure 6.2 Overall timeline for the standardization of IMT-2020.

6.4 Overview of IMT-2020

It is envisaged that IMT-2020 will expand and support diverse usage scenarios and applications that will extend beyond those currently supported by IMT systems. A broad variety of capabilities will need to be tightly coupled with these intended different usage scenarios and applications for IMT-2020.

6.4.1 Usage Scenarios of IMT-2020

Three main usage scenarios for IMT-2020 have been identified in Recommendation ITU-R M.2083, "IMT Vision – Framework and overall objectives of the future development of IMT for 2020 and beyond", which are enhanced mobile broadband, ultra-reliable and low latency communications, and massive machine-type communications. Additional use cases are expected to emerge, which are currently not foreseen. For future IMT, flexibility will be necessary to adapt to new use cases that come with a widely varying range of requirements.

IMT-2020 will encompass a large number of different features. Depending on the circumstances and the different needs in different countries, future IMT

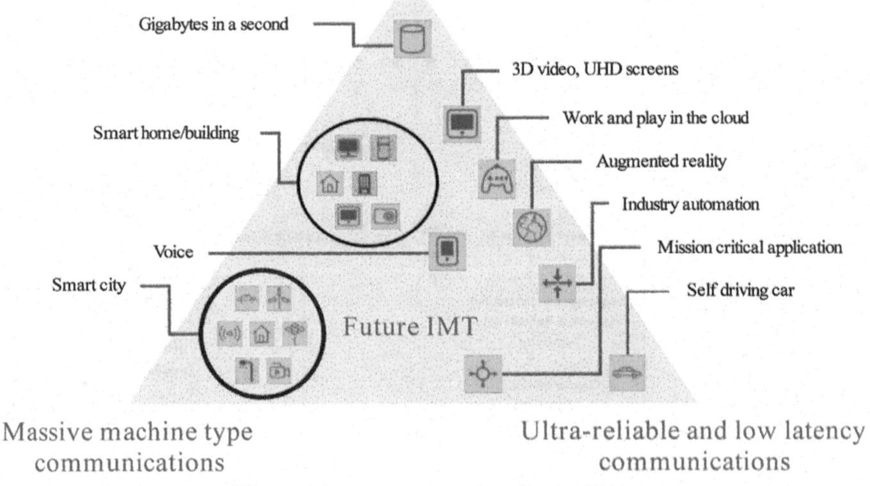

Figure 6.3 Usage scenarios of IMT-2020.

systems should be designed in a highly modular manner so that not all features have to be implemented in all networks.

Figure 6.3 illustrates some examples of envisioned usage scenarios for IMT-2020.

6.4.2 Capabilities of IMT-2020

IMT-2020 systems are mobile systems that include the new capabilities of IMT that go beyond those of IMT-Advanced. IMT-2020 systems will need to support low to high mobility applications and a wide range of data rates in accordance with user and service demands in multiple usage scenarios. IMT-2020 also needs to have capabilities for high quality multimedia applications within a wide range of services and platforms, providing a significant improvement in performance and quality of service. The key design principles of capabilities of IMT-2020 are flexibility and diversity to serve many different use cases and scenarios.

The following eight parameters are considered to be key capabilities of IMT-2020:

Peak data rate

Maximum achievable data rate under ideal conditions per user/device (in Gbit/s).

User experienced data rate

Achievable data rate that is available ubiquitously across the coverage area to a mobile user/device (in Mbit/s or Gbit/s).

Latency

The contribution by the radio network to the time from when the source sends a packet to when the destination receives it (in ms).

Mobility

Maximum speed at which a defined QoS and seamless transfer between radio nodes which may belong to different layers and/or radio access technologies (multi-layer/-RAT) can be achieved (in km/h).

Connection density

Total number of connected and/or accessible devices per unit area (per km^2).

Energy efficiency

Energy efficiency has two aspects:

- on the network side, energy efficiency refers to the quantity of information bits transmitted to/received from users, per unit of energy consumption of the radio access network (RAN) (in bit/Joule);
- on the device side, energy efficiency refers to quantity of information bits per unit of energy consumption of the communication module (in bit/Joule).

Spectrum efficiency

Average data throughput per unit of spectrum resource and per cell (bit/s/Hz).

Area traffic capacity

Total traffic throughput served per geographic area (in Mbit/s/m^2).

The key capabilities of IMT-2020 are shown in Figure 6.4, compared with those of IMT Advanced.

6.5 Key Technology Enablers

Report ITU-R M.2320 [7] provides a broad view of future technical aspects of terrestrial IMT systems considering the timeframe 2015–2020 and beyond. It includes information on technical and operational characteristics of IMT systems, including the evolution of IMT through advances in technology and

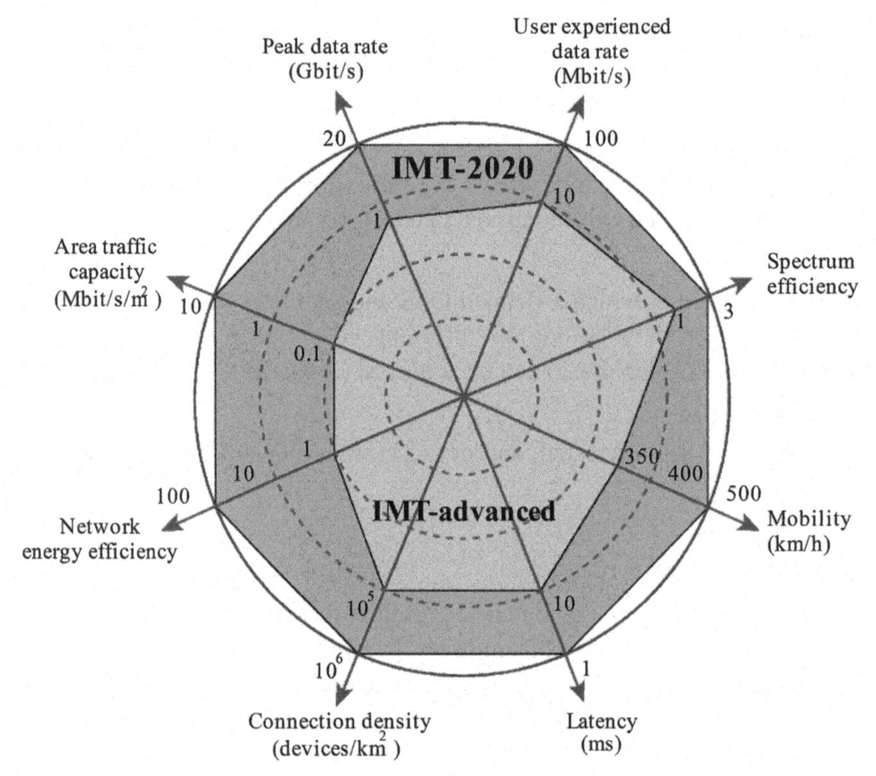

Figure 6.4 Enhancement of key capabilities from IMT-Advanced to IMT-2020.

spectrally-efficient techniques, and their deployment. Key technologies that are expected to influence the development of IMT-2020 are briefly described below.

6.5.1 Technologies to Enhance the Radio Interface

Advanced waveforms, modulation and coding, and multiple access schemes, e.g., filtered OFDM (FOFDM), filter bank multi-carrier modulation (FBMC), pattern division multiple access (PDMA), sparse code multiple access (SCMA), interleave division multiple access (IDMA) and low density spreading (LDS) may improve the spectral efficiency of the future IMT systems.

Advanced antenna technologies such as 3D-beamforming (3D-BF), active antenna system (AAS), massive MIMO and network MIMO will achieve better spectrum efficiency.

In addition, TDD-FDD joint operation, dual connectivity and dynamic TDD can enhance the spectrum flexibility.

Simultaneous transmission and reception on the same frequency with self-interference cancellation could also increase spectrum efficiency.

Other techniques such as flexible backhaul and dynamic radio access configurations can also enable enhancements to the radio interface.

In small cells, higher-order modulation and modifications to the reference-signal structure with reduced overhead may provide performance enhancements due to lower mobility in small cell deployments and potentially higher signal-to interference ratios compared to the wide-area case.

Flexible spectrum usage, joint management of multiple radio access technologies (RATs) and flexible uplink/downlink resource allocation, can provide technical solutions to address the growing traffic demand in the future and may allow more efficient use of radio resources.

6.5.2 Network Technologies

Future IMT will require more flexible network nodes which are configurable based on the Software-Defined Networking (SDN) architecture and network function virtualization (NFV) for optimal processing the node functions and improving the operational efficiency of network.

Featuring centralized and collaborative system operation, the cloud RAN (C-RAN) encompasses the baseband and higher layer processing resources to form a pool so that these resources can be managed and allocated dynamically on demand, while the radio units and antenna are deployed in a distributed manner.

The radio access network (RAN) architecture should support a wide range of options for inter-cell coordination schemes. The advanced self organizing network (SON) technology is one example solution to enable operators to improve the OPEX efficiency of the multi-RAT and multi-layer network, while satisfying the increasing throughput requirement of subscribers.

6.5.3 Technologies to Enhance Mobile Broadband Scenarios

A relay based multi-hop network can greatly enhance the Quality of Service (QoS) of cell edge users. Small-cell deployment can improve the QoS of users by decreasing the number of users in a cell and user quality of experience can be enhanced.

Dynamic adaptive streaming over HTTP (DASH) enhancement is expected to improve user experience and accommodate more video streaming content in existing infrastructure.

Bandwidth saving and transmission efficiency improvement is an evolving trend for Evolved Multimedia Broadcast and Multicast Service (eMBMS). Dynamic switching between unicast and multicast transmission can be beneficial.

IMT systems currently provide support for RLAN interworking, at the core network level, including seamless as well as non-seamless mobility, and can offload traffic from cellular networks into license-exempt spectrum bands.

Context aware applications may provide more personalized services that ensure high QoE for the end user and proactive adaptation to the changing context.

Proximity-based techniques can provide applications with information whether two devices are in close proximity of each other, as well as enabling direct device-to-device (D2D) communication. Group communication, including push-to-talk type of communication, is highly desirable for public safety applications.

6.5.4 Technologies to Enhance Massive Machine Type Communications

Future IMT systems are expected to connect a large number of M2M devices with a range of performance and operational requirements, with further improvement of low-cost and low complexity device types as well as extension of coverage.

6.5.5 Technologies to Enhance Ultra-reliable and Low Latency Communications

To achieve ultra-low latency, the data and control planes may both require significant enhancements and new technical solutions addressing both the radio interface and network architecture aspects.

It is envisioned that future wireless systems will, to a larger extent, also be used in the context of machine-to-machine communications, for instance in the field of traffic safety, traffic efficiency, smart grid, e-health, wireless industry automation, augmented reality, remote tactile control and tele-protection, requiring high reliability techniques.

6.5.6 Technologies to Improve Network Energy Efficiency

In order to enhance energy efficiency, energy consumption should be considered in the protocol design.

The energy efficiency of a network can be improved by both reducing RF transmit power and saving circuit power. To enhance energy efficiency, the traffic variation characteristic of different users should be well exploited for adaptive resource management. Examples include discontinuous transmission (DTX), base station and antenna muting, and traffic balancing among multiple RATs.

6.5.7 Terminal Technologies

The mobile terminal will become a more human friendly companion as a multipurpose Information and Communication Technology (ICT) device for personal office and entertainment, and will also evolve from being predominantly a hand-held smart phone to also include wearable smart devices.

Technologies for chip, battery, and display should therefore be further improved.

6.5.8 Technologies to Enhance Privacy and Security

Future IMT systems need to provide robust and secure solutions to counter the threats to security and privacy brought by new radio technologies, new services and new deployment cases.

6.6 Spectrum for IMT Operation

This Section discusses about needs for IMT operation spectrum as well as harmonization, identification and technical issues on spectrum for IMT.

6.6.1 Spectrum Requirements

Report ITU-R M.2290 [8] provides the results of studies on estimated global spectrum requirements for terrestrial IMT in the year 2020. The estimated total requirements include spectrum already identified for IMT plus additional spectrum requirements.

It is noted that no single frequency range satisfies all the criteria required to deploy IMT systems, particularly in countries with diverse geographic and population density; therefore, to meet the capacity and coverage requirements of IMT systems multiple frequency ranges would be needed. It should be noted that there are differences in the markets and deployments and timings of the mobile data growth in different countries.

For future IMT systems in the year 2020 and beyond, contiguous and broader channel bandwidths than available to current IMT systems would be desirable to support continued growth. Therefore, availability of spectrum resources that could support broader, contiguous channel bandwidths in this time frame should be explored.

6.6.2 Studies on Technical Feasibility of IMT between 6 and 100 GHz

Report ITU-R M.2376 [9] provides information on the technical feasibility of IMT in the frequencies between 6 and 100 GHz. It includes information on potential new IMT radio technologies and system approaches, which could be appropriate for operation in this frequency range. The potential advantages of using the same spectrum for both access and fronthaul/backhaul, as compared with using two different frequencies for access and fronthaul/backhaul, are also described in the Report. The theoretical assessment, simulations, measurements, technology development and prototyping described in the Report indicate that utilizing the bands between 6 and 100 GHz is feasible for studied IMT deployment scenarios, and could be considered for the development of IMT for 2020 and beyond.

6.6.3 Spectrum Harmonization

Where radio systems are to be used globally, it is highly desirable for existing and newly allocated spectrum to be harmonized. The benefits of spectrum harmonization include: facilitating economies of scale, enabling global roaming, reducing equipment design complexity, preserving battery life, improving spectrum efficiency and potentially reducing cross border interference.

Mobile devices typically contain multiple antennas and associated radio frequency front-ends to enable operation in multiple bands to facilitate roaming. While mobile devices can benefit from common chipsets, variances in frequency arrangements necessitate different components to accommodate these differences, which leads to higher equipment design complexity.

Consequently, harmonization of spectrum for IMT will lead to simplification and commonality of equipment, which is desirable for achieving economies of scale and affordability of equipment.

6.6.4 Spectrum Identification

As mentioned previously, it was by a decision by the 1992 World Administrative Radiocommunication Conference that the first specific frequency bands for the operation of FPLMTS (now IMT) were identified in the ITU Radio

Regulations, the international treaty governing the use of the radio frequency spectrum and satellite orbits. Identification of a frequency band in the Radio Regulations does not afford any priority for such use with respect to other radio services allocated to that spectrum, but it does provide a clear signal to the national regulators for their spectrum planning, and also provides a degree of confidence for equipment manufacturers and network operators to make the long term investments necessary to develop IMT in these bands.

Since then, successive World Radiocommunication Conferences have periodically identified additional frequency bands for IMT within the range of 450 MHz to 6 GHz to cater for the rapidly growing demand for mobile communications, particularly mobile broadband data.

While the 2015 World Radiocommunication Conference made good progress in identifying additional frequency bands and globally harmonized arrangements below 6 GHz for the operation of IMT, it also recognized a potential future requirement for large contiguous blocks of spectrum at higher frequencies for these systems.

Consequently, it called for 11 frequency bands above 24 GHz to be studied by ITU-R as bands that may be identified for use by IMT at the World Radiocommunication Conference in 2019 (WRC-19). The frequency bands to be studied are shown in Table 6.1 below.

As can be seen, the different bands span from around 24 GHz up to 86 GHz. While some of those bands are already allocated for the operation of mobile services in the Radio Regulations, others would require a mobile service allocation in addition to an identification for operation of IMT systems.

The focus of these studies should be to identify a limited subset of these bands that are recommended to be identified globally for use by IMT. The studies on these bands will be conducted in a Task Group of ITU-R Study Group 5, and the results of the studies will be included in the Conference Preparatory Meeting report to the World Radiocommunication Conference 2019.

Table 6.1 Bands under study for IMT identification by WRC-19

Existing Mobile Allocation	No Global Mobile Allocation
24.25 GHz–27.5 GHz	31.8–33.4 GHz
37–40.5 GHz	40.5 –42.5 GHz
42.5–43.5 GHz	
45.5–47 GHz	47–47.2 GHz
47.2–50.2 GHz	
50.4–52.6 GHz	
66–76 GHz	
81–86 GHz	

6.7 Conclusions

The scope of IMT-2020 is much broader than the previous generations of mobile broadband communication systems. We are talking here about not just an enhancement to the traditional mobile broadband scenarios, but extending the application of this technology to use cases involving ultra-reliable and low latency communications, and massive machine-type communications. The ITU's work in developing the specifications for IMT-2020, in close collaboration with the whole gamut of 5G stakeholders, is now well underway, along with the associated spectrum management and spectrum identification aspects.

IMT is the enabler of new trends in communication devices – from the connected car and intelligent transport systems to augmented reality, holography, and wearable devices, and a key enabler to meet social needs in the areas of mobile education, connected health and emergency telecommunications. E-applications are transforming the way we do business and govern our countries, and smart cities are pointing the way to cleaner, safer, more comfortable lives in our urban conglomerates. Certainly, IMT-2020 will be a cornerstone for all of the activities related to attaining the goals in the 2030 Agenda for Sustainable Development.

References

[1] Recommendation ITU-R M.1457 – "Detailed specifications of the terrestrial radio interfaces of International Mobile Telecommunications-2000 (IMT-2000)", May 2000, http://www.itu.int/rec/R-REC-M.1457
[2] ITU-R Resolution 56 – "Naming for International Mobile Telecommunications", October 2007, http://www.itu.int/pub/R-RES-R.56
[3] ITU-R Resolution 57 – "Principles for the process of development of IMT-Advanced", October 2007, http://www.itu.int/pub/R-RES-R.57
[4] Recommendation ITU-R M.2012 – "Detailed specifications of the terrestrial radio interfaces of International Mobile Telecommunications Advanced (IMT-Advanced)", January 2012, http://www.itu.int/rec/R-REC-M.2012
[5] Recommendation ITU-R M.2083 – "IMT Vision – Framework and overall objectives of the future development of IMT for 2020 and beyond", September 2015, http://www.itu.int/rec/R-REC-M.2083
[6] Circular Letter 5/LCCE/59 – "Invitation for submission of proposals for candidate radio interface technologies for the terrestrial components

of the radio interface(s) for IMT-2020 and invitation to participate in their subsequent evaluation", 22 March 2016, http://www.itu.int/md/R00-SG05-CIR-0059/
[7] Report ITU-R M.2320 – "Future technology trends of terrestrial IMT systems", 2014, http://www.itu.int/pub/R-REP-M.2320
[8] Report ITU-R M.2290 – "Future spectrum requirements estimate for terrestrial IMT", December 2013, http://www.itu.int/pub/R-REP-M.2290
[9] Report ITU-R M.2376 – "Technical feasibility of IMT in bands above 6 GHz", July 2015, http://www.itu.int/pub/R-REP-M.2376

About the Author

François Rancy was elected Director of the ITU Radiocommunication Bureau by the ITU Plenipotentiary Conference 2010 (PP-10) in Guadalajara, Mexico. He took office on 1 January 2011. He was confirmed in a second term by the ITU Plenipotentiary Conference 2014 (PP-14) in Busan, Republic of Korea.

François Rancy is an engineer, a graduate of the École Polytechnique (1977) and the École Nationale Supérieure des Télécommunications (Paris, 1979).

From 1979 to 1997, he worked as systems engineer and subsequently Head of Department in France Télécom's research laboratories, where he was in charge of studies on national and international satellite systems and activities relating to the spectrum and the regulation of satellite systems. As from 1992, his responsibilities expanded to cover the entire radiocommunication sphere.

From 1997 to 2004, he was Director of Spectrum Planning and International Affairs at the National Frequency Agency (ANFR).

From 2004 to 2010, he was Director-General of ANFR, responsible for frequency management in France.

In October 2010, he was elected Director, BR by the ITU Plenipotentiary Conference.

At the international level, he led the coordination of European delegations (CEPT) at WRC-03 and chaired the European Union group on spectrum management policy (RSPG) in 2007. He chaired the ITU–R Special Committee on Regulatory and Procedural Matters from 1997 to 2003, and the World Radiocommunication Conference in 2007.

François Rancy has been awarded the ITU Silver Medal (2007) and the titles of Chevalier de la Légion d'honneur (1998) and Ordre national du mérite (1992).

7

Connectivity of Ad hoc 5G Wireless Networks under Denial of Service Attacks

Zituo Jin[1], Santhanakrishnan Anand[2], Koduvayur P. Subbalakshmi[3] and Rajarathnam Chandramouli[3]

[1]Courant Institute of Mathematical Sciences, New York University, USA
[2]Department of Electrical Engineering and Computer Science, New York Institute of Technology, USA
[3]Department of Electrical and Computer Engineering, Stevens Institute of Technology, USA

7.1 Introduction

This chapter studies the network layer effects of denial of service (DoS) in 5G wireless networks. DoS attacks by malicious nodes have mostly been researched in terms of their effect on the physical layer (i.e., loss of wireless connectivity) for the good nodes or in terms of loss of services for the good nodes. However, one vital consequence of DoS on the network layer is that it can cause unreliable, disconnected networks. This chapter presents an analysis to study the connectivity of ad hoc 5G networks under DoS. The ad hoc network is modeled as a random geometric graph and an approximation for the probability that the network is disconnected due to DoS is determined. This research indicates that in the absence of suitable defense mechanisms, DoS can disconnect a network with significantly high probability. Then, alleviation of this effect by cooperation among good users is discussed. For this, spectrum decision protocols are analyzed, in which good users make individual spectrum decisions to detect DoS and then exchange individual sensing results with their one-hop neighbors (i.e., a distributed protocol) or with a centralized controller (a centralized protocol) to increase resilience to DoS. Our distributed protocol can reduce the probability of the network becoming disconnected under DoS by 31% to two orders of magnitude. The centralized protocol can almost eliminate the effect of DoS on the network layer under certain spatial density of malicious users.

The number of connected Internet devices is expected to reach 50 billion by the year 2020 [1]. The Fifth generation (5G) wireless networks vision is to provide connectivity with low latency, enhanced quality-of-service (QoS), low energy, mobility, large spectral efficiency for all the devices taking into account the growth in the number of devices [2]. This motivated the deployment of unlicensed access to the 4G long term evolution (LTE) networks [3]. In the recent AWS-3 spectrum auction, cellular companies spent $44.9 billion for licenses to 65 MHz of spectrum [4]. Though unlicensed spectrum is less valuable to cellular companies due to stricter transmission rules and the inability to limit other interferers, the potential to use such bands to augment capacity in certain areas is attractive because devices can use these bands at no cost. For example, in 2014, up to 46% of potential cellular data was offloaded to Wi-Fi [5].

Dynamic spectrum access (DSA) [6] based cognitive radio networks [7] are expected to be an integral part of 5G wireless networks. Device-to-device (D2D) communication is also recognized as one of the technology components of the evolving 5G architecture by the European Union project, Mobile and wireless communications Enablers for the Twenty-twenty Information Society (METIS) [8]. Integration of D2D communication into incumbent 3GPP networks was discussed in [9]. Ye et al. [10] compared infrastructure vs ad hoc mode of communication for D2D communications and found ad hoc mode to be better. More detailed analysis for the achievable rates and signal-to-interference-noise-ratio (SINR) distributions were provided in [11]. Since D2D communications take place without infrastructure, it is essential to efficiently design ad hoc 5G wireless networks.

The access of LTE devices to unlicensed spectrum (or D2D communications) is achieved by "listen before talk (LBT)" spectrum etiquette [3] similar to the secondary user spectrum etiquette in dynamic spectrum access (DSA) networks [12]. Al-Dulaimi et al. proposed a modified LBT etiquette to take into account the interference constraints on Wi-Fi users [13]. However, the access to unrestricted unlicensed spectrum or D2D communications can be susceptible to denial of service (DoS) attacks [14], which have not been discussed in detail for 5G wireless networks. Few types of DoS attacks and their mitigation are discussed in [15, 16]. Li et al. discuss denial of service attacks and malicious attacks [15], where in DoS is measured based on loss of QoS and frequencies that experience DoS are black-listed. Such mechanisms can end up being inadequate because it results in false alarms although the probability of detection is large [17]. In [16], Klassen and Yang perform NS-3 based simulations to show that DoS attack can be mitigated by intrusion detection. However, these studies are not applicable to 5G wireless networks because in unlicensed access, malicious nodes can neither be localized nor identified.

The DoS attack that 5G networks are susceptible to, is similar to primary user emulation attack (PUEA), introduced in [18, 19], unique to cognitive radio enabled dynamic spectrum access (DSA) networks [20]. In this type of attack, a set of malicious secondary users mimic the primary transmitter, leading other secondary users to believe that a primary user is active, when, in fact, it is not. This makes the good (non-malicious) secondary users, following normal spectrum evacuation process, to vacate the spectrum unnecessarily, causing spectrum wastage. PUEA was first discussed by Chen et al. in [18] and [19]. In [18], two mechanisms to detect PUEA were proposed, namely, the distance ratio test and the distance difference test based on the correlation between the length of wireless link and the received signal strength. In [19], a defense mechanism against PUEA was proposed by locating the spurious transmission via an underlying sensor network and comparing it with the known location of the primary transmitter. Anand et al. presented the first analytical model to characterize the probability of successful PUEA based on energy detection in [21]. Jin et al. then proposed two hypothesis test based approaches [22, 23], which enable each individual good secondary user to detect PUEA.

In the literature, PUEA or DoS in general, has mostly been studied from the perspective of its impact on the physical (PHY) layer, but DoS can also affect the performance of DSA networks at higher layers.

As an example, at the link layer, DoS can cause blocking of new calls and dropping of ongoing calls for delay intolerant traffic and cause additional delay for delay tolerant traffic. Jin et al. presented the first analysis of link layer effects of DoS attacks in [24] and [25], where it was shown how PUEA can adversely affect the link layer performance if it is not carefully addressed. DoS also affects connectivity of the network. Xing and Wang discussed connectivity of ad hoc networks subject to DoS attacks [26]. However, they only analyzed nodes that get isolated (i.e., have no links incident on them). As will be shown in the example in Section 7.2, DoS can disconnect a networks without necessarily creating isolated nodes. Therefore, a more detailed analysis of the connectivity of ad hoc networks under DoS is required.

This chapter, addresses two issues regarding the network level impact of DoS, i.e., (i) Can DoS also affect the network connectivity of ad hoc 5G wireless networks? and (ii) Can good users collaborate in a centralized and distributed manner, to mitigate the adverse effect of DoS on the network layer?

First, the ad hoc network is modeled as a random geometric graph and an Erdös-Rényi graph based approximate analysis is performed to compute the probability that DoS disconnects the network. Then, centralized [27] and distributed [28] cooperation between the good users is studied to mitigate this effect of DoS. Results indicate that for a small network (a network with 100 secondary users or less), a 5% probability of successful DoS attack results in a significantly large (more than 20%) probability of disconnecting a connected ad hoc network. For small number of malicious users, the centralized protocol proposed in [27] can almost surely guarantee a connected network even in the presence of DoS. For large number of malicious users, the centralized protocol can reduce the probability of the network being disconnected as a consequence of DoS, by about 20%. The distributed protocol presented in [28], can reduce the probability of DoS causing a disconnected network, by about 30% to two orders of magnitude.

7.2 Problem Definition

Consider an example ad-hoc network with three available channels, as shown in Figure 7.1. Nodes communicate with each other using one of three spectrum bands. Figure 7.1 shows a snapshot in which node-pairs (A, D), (B, C) and (E, F) communicate on channel 1.

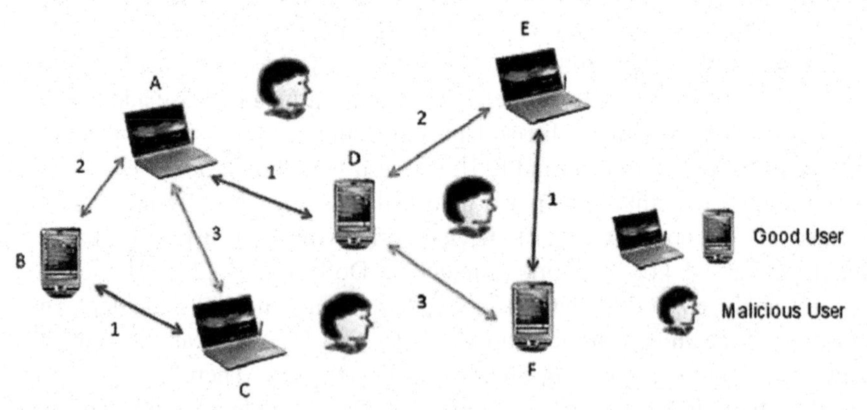

Figure 7.1 A typical ad-hoc network. Nodes A and D are cut-vertices. If DoS is successfully launched on channel 1, node-pairs (A, D), (B, C) and (E, F) will vacate channel 1, thereby resulting in a disconnected network.

If DoS is successfully launched on channel 1, then these node-pairs need to look for another channel for communication. Since all the three available channels are already in use, these node-pairs cannot find an alternate channel and hence lose the connectivity between them. Since A and D form the cut-vertices of the network, the set of nodes, {A, B, C} will find no means of reaching the set of nodes, {D, E, F}, thus rendering the network disconnected and hence, unreliable. Based on this example, it can be argued that DoS can affect the network layer performance of ad hoc networks adversely, if it is not carefully mitigated.

Since the primary function of the network layer is the routing of data packets to their destinations, a basic requirement for efficient network layer performance is the connectivity of the underlying ad hoc network graph. The good users form the vertices of the underlying network graph and two vertices share an edge if the corresponding secondary users are one hop neighbors, i.e., they are within a specified radio "hearing distance", R, from each other. The network itself is connected if it is possible for packets from any node to reach any other node (possibly over several hops). One metric that can be used to measure the network layer performance is the probability of the network being connected.

If malicious users launch a DoS attack, some of the good users may succumb to the attack and evacuate the spectrum band, while some may be able to detect the attack and stay in the band. As shown in Figure 7.2, the set of remaining nodes could possibly remain connected to each other or could also form a disconnected network where at least two nodes cannot reach each other. The probability that the network becomes disconnected under DoS, is used as a metric to gauge the severity of the attack. The analysis for the case when the network is connected before DoS is launched, is presented. The analysis can easily be extended to the case when the network is disconnected before DoS, by studying each connected component separately.

The objectives in this chapter are as follows.

- Develop an analysis to measure the probability that DoS disconnects a connected ad hoc network.
- Study the probability that DoS renders a connected ad hoc network, disconnected, when deploying the centralized protocol developed in [27] and the distributed protocol developed in [28] are deployed.

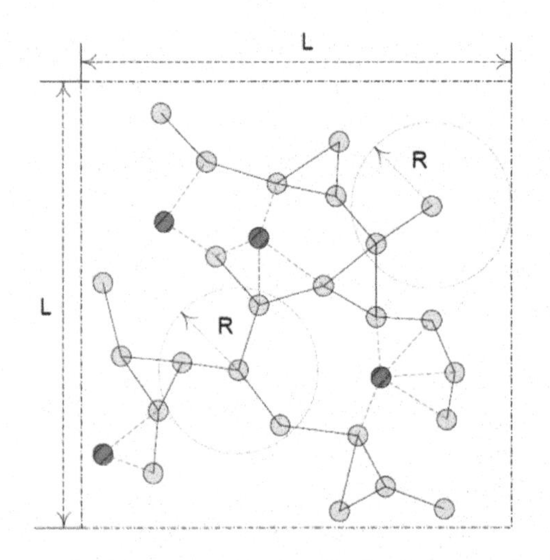

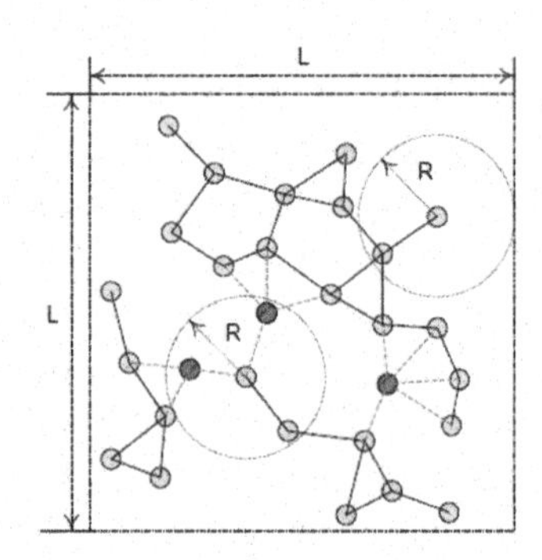

(a) Connected network after DoS attack (b) Disconnected network after DoS attack

Figure 7.2 A connected ad hoc network with users independently and uniformly distributed in an L × L grid. All users have "hearing distance" R, within which they can communicate with each other. Each user could potentially be affected by DoS. The users succumb to the attack are marked red and have to be removed, along with all associated links. The removal of the DoS victims could lead to either a connected or disconnected network.

7.3 Connectivity Analysis

The underlying network graph can be modeled as a random geometric graph [29], where each vertex represents a good user and an edge between two vertices indicates that the two corresponding users are within a distance, *R*, of each other. The DSA network can be viewed as a random geometric graph $G(N, p_{ij})$, where *N* denotes the number of vertices (users), and p_{ij} denotes the probability that there exists an edge between the *i*th and the *j*th vertices, i.e., the probability that vertices *i* and *j* are less than a distance *R* apart. Note that $p_{ij} = p_{ji}, \forall\, i, j$.

The probability, $\mathrm{p_{ij}}$, is computed as follows. Let node i be located at $(\mathrm{x_i}, \mathrm{y_i})$ and node j be located at $(\mathrm{x_j}, \mathrm{y_j})$. The co-ordinates, $\mathrm{x_i}$, $\mathrm{y_i}$, $\mathrm{x_j}$ and $\mathrm{y_j}$ are all independent and uniformly distributed in $(0, \mathrm{L})$. Conditioned on $\mathrm{x_i}$ and $\mathrm{y_i}$, the probability, $\mathrm{p_{ij}}(\mathrm{x_i}, \mathrm{y_i})$ can be written as

$$p_{ij}(x_i, y_i) = \frac{1}{L^2} \int_{y_j = y_{\min}}^{y_{\max}} \int_{x_j = x_{\min}}^{x_{\max}} dx_j dy_j, \qquad (7.1)$$

where
$x_{min} = \max(0, x_i - \sqrt{R^2 - (y_i - y_j)^2})$, $x_{max} = \min(L, x_i + \sqrt{R^2 - (y_i - y_j)^2})$, $y_{min} = \max(0, y_i - R)$ and $y_{max} = \min(L, y_i + R)$. Averaging over x_i and y_i, p_{ij} is obtained as

$$p_{ij} = \frac{1}{L^2} \int_{y_i=0}^{L} \int_{x_i=0}^{L} p_{ij}(x_i, y_i) dx_i dy_i, \tag{7.2}$$

which is independent of i and j since x_i, y_i, x_j and y_j are all mutually independent and identically distributed. Therefore, p_{ij} in Equation (7.2) can be written as $\overline{p}$ and the random geometric graph can be represented as $G(N, \overline{p})$. First, the probability that $G(N, \overline{p})$ is connected needs to be computed. Since the analysis of random geometric graphs is complex because some links are dependent on others due to *triangle inequality* [29], the graph is analyzed as if it were a Erdös-Rényi graph [30][1].

Let $p_c(N)$ denote the probability that $G(N, \overline{p})$ is connected and $q(N) \triangleq 1 - p_c(N)$ denote the probability that $G(N, \overline{p})$ is disconnected. When $G(N, \overline{p})$ is disconnected, any arbitrary vertex, i, must belong to a component with k vertices, $1 \leq k \leq N-1$. Let $G(k, \overline{p})$ denote the sub-graph that contains the arbitrary vertex, i. An arbitrary vertex, i, belongs to a component with exactly, k vertices, with probability, $p^{(i)}(k)$, when $G(k, \overline{p})$ is connected and there is no edge from any vertex in $G(k, \overline{p})$ to any of the remaining $N - k$ vertices. Therefore, $p^{(i)}(k)$ can be written as

$$p^{(i)}(k) = p_c(k)(1 - \overline{p})^{k(N-k)} \tag{7.3}$$

The probability that $G(N, \overline{p})$ is initially disconnected, $q(N)$, can then be written as

$$q(N) = \sum_{k=1}^{N-1} \binom{N-1}{k-1} p_c(k)(1 - \overline{p})^{k(N-k)}, \tag{7.4}$$

and hence, the probability that $G(N, \overline{p})$ is initially connected, $p_c(N)$, can then be written as

$$\begin{aligned} p_c(N) &= 1 - q(N) \\ &= 1 - \sum_{k=1}^{N-1} \binom{N-1}{k-1} p_c(k)(1 - \overline{p})^{k(N-k)}, \end{aligned} \tag{7.5}$$

where $p_c(1) = 1$ and $p_c(2) = \overline{p}$.

[1]Such approximations have been done in the past, e.g., [31].

Now, it is essential to calculate the probability that DoS disconnects an initially connected network. All nodes succumb to DoS with probability, p_{DoS} independent of each other. When a node is affected by DoS, the node and all edges incident on the node have to be removed. This is equivalent to independently and randomly removing vertices in $\mathrm{G}(\mathrm{N}, \overline{\mathrm{p}})$, along with all associated links, with probability p_{DoS}. There moval, thus, can be modeled as a Bernoulli trial with success probability p_{DoS}. Hence, the number of secondary users becoming victims of DoS, n_{DoS}, is binomially distributed, i.e., $n_{\text{DoS}} \sim Binomial\ (N, p_{\text{DoS}})$. After the removal process is completed, the resultant graph is modeled as a new random Erdös-Rényi graph $G(\widehat{N},\ \acute{\overline{p}})$, where $\acute{\overline{p}} = \overline{p}$ and $\widehat{N}$ is given by

$$\widehat{N} = N - E[n_{\text{DoS}}] = N(1 - p_{\text{DoS}}). \tag{7.6}$$

The probability that $G(\widehat{\mathrm{N}}, \overline{p})$ is connected, $p_c(\widehat{N})$, can be obtained from Equation (7.5). The probability that DoS disconnects a network that is initially connected, i.e., the conditional probability, $\mathrm{P_r}\{G(\acute{N}, \overline{p})\ \text{disconnected}|$ $G(N, \overline{p})\text{connected}\}$, denoted by $p_{\text{disconnect}}$, can be written as

$$p_{\text{disconnect}} = 1 - \frac{\mathrm{P_r}\{G(\widehat{N}, \overline{p})\text{connected}, G(N, \overline{p})\text{connected}\}}{\mathrm{P_r}\{G\ (N, \overline{p})\text{connected}\}} \tag{7.7}$$

Since the event that $G(\hat{N}, \overline{p})$ is connected is a subset of the event that $G(N, \overline{p})$ is connected, $p_{\text{disconnect}}$ can be evaluated as

$$p_{\text{disconnect}} = 1 - \frac{\mathrm{P_r}\{G(\widehat{N}, \overline{p})\text{connected}\}}{\mathrm{P_r}\ \{G(N, \overline{p})\text{connected}\}} = 1 - \frac{p_c(\widehat{N})}{p_c(N)} \tag{7.8}$$

where $p_c(.)$ is given by Equation (7.5).

The discussion thus far for the derivation of $p_{\text{disconnect}}$ in Equation (7.8) does not consider any defense mechanisms against DoS. It is noted from Equation (7.8) that the factor that can be controlled to reduce the value of $p_{\text{disconnect}}$ is the probability that nodes succumb to PUEA, p_{DoS}. One can reduce p_{DoS} by deploying the detection mechanism for each individual secondary user, specified in [27]. The value of p_{DoS} can be further reduced by deploying the centralized protocol presented in [27] or the distributed protocol developed in [28]. The centralized protocol developed in [27] is more efficient in keeping the network connected even in the presence of DoS attacks (as will be seen in Section 7.4), but requires the presence of a centralized controller, which may or may not be possible in all DSA networks. The distributed protocol is not as efficient as the centralized protocol (as will be observed from the results depicted in Section 7.4), but is less complex to implement.

7.4 Results and Discussions

The locations of all the users including both good and malicious users are considered to be uniformly distributed in a 2000 m $\times$ 2000 m square grid (i.e., $L = 2000$). Each user has a transmission range of $R = 250$ m [7]. Plugging in these values in Equations (7.1) and (7.2), $\overline{p} = 0.0436$. The number of users in the network (i.e., the number of vertices in $G(N, \overline{p})$), N, is set as 100, 200 and 500, respectively. First, the effect of DoS on the connectivity of the network is evaluated. For this, the probability of successful DoS attack, p_{DoS}, is varied from 0 to 1 in increments of 0.05. Figure 7.3(a) depicts the probabilities that DoS disconnects an initially connected ad hoc network (i.e., $p_{\mathrm{disconnect}}$ specified in Equation (7.8)) with different numbers of users. The legends in Figure 7.3(a) are explained as follows. The numbers “100”, “200” and “500” represent 100, 200 and 500 secondary users in the network, respectively, while the parenthesized letters “S” and “A” represent simulations results (obtained by C based simulations on UBUNTU Linux platform) and analytical results (using the analysis presented in Section 7.3), respectively.

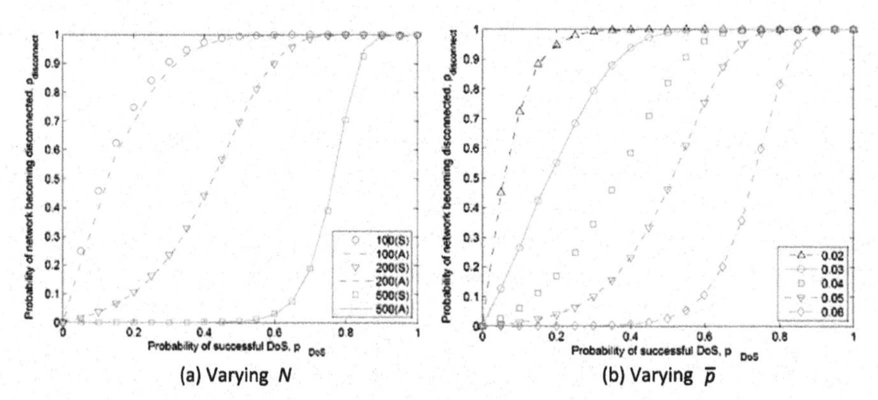

Figure 7.3 Comparison of probability that PUEA disconnects an initially connected network with different numbers of secondary users, N and different values of the probability that any two secondary users have a link, $\overline{p}$. In Figure (a), $\overline{p} = 0.0436$. The numbers “100”, “200” and “500” in the legends represent 100, 200 and 500 secondary users in the network, respectively, while the parenthesized letters “S” and “A” represent the results from simulations and analysis, respectively. In Figure (b), The numbers “0.02”, “0.03”, “0.04”, “0.05” and “0.06” in the legends represent the probabilities that any two secondary users have a direct link set as “0.02”, “0.03”, “0.04”, “0.05” and “0.06”, respectively.

It can be seen from Figure 7.3(a) that as the number of users increases, the network becomes more resilient to disconnectedness under DoS. For example, when $p_{\text{DoS}} = 0.45$, the networks with 100 and 200 secondary users can have a $p_{\text{disconnect}}$ up to about 0.99 and 0.57, respectively, while the network with 500 secondary users achieves almost 0 for $p_{\text{disconnect}}$. That is equivalent to saying that the network with 500 secondary users would almost surely remain connected under DoS as long as the probability of successful DoS attack does not exceed 0.45. This is because, as N increases, for the same R, i.e., for the same value of $\overline{p}$, the average number of neighbors, $N\overline{p}$, increases for each node. This means that the degree of each node increases and therefore, the probability that the network is connected, increases [29, 30]. It is also observed from Figure 7.3(a) that when the probability of successful DoS attack is high enough, e.g., when $p_{\text{DoS}} = 0.9$, which is the case in networks with no defense mechanisms for DoS, even the network with 500 secondary users becomes disconnected under DoS.

The effect of the probability that any two secondary users have a direct link, $\overline{p}$, on the probability that DoS disconnects a initially connected network, $p_{\text{disconnect}}$, is also studied. The value of $\overline{p}$ can be varied in practice, either by modifying R or by increasing the area of the square grid, $L2$. The number of secondary users is fixed at $N = 200$ and the probability of successful PUEA p_{DoS} is varied from 0 to 1 in increments of 0.05. Figure 7.3(b) depicts the probabilities that DoS disconnects an initially connected network with respect to p_{DoS} for varying values of $\overline{p}$. The numbers "0.02", "0.03", "0.04", "0.05" and "0.06" in the legends represent $\overline{p} =$ "0.02", "0.03", "0.04", "0.05" and "0.06", respectively. It is observed from Figure 7.3(b) that as $\overline{p}$ increases from 0.02 to 0.06, $p_{\text{disconnect}}$ decreases, indicating that the network is more likely to remain connected under DoS. For example, when $p_{\text{DoS}} = 0.4$ and $\overline{p} = 0.02$, the network is almost surely disconnected, while for the same value of p_{DoS}, when $\overline{p} = 0.06$, the network is almost surely connected. This is intuitively correct because $\overline{p}$ increases when the grid size, L, decreases or when the transmission range, R, increases. In both these cases, more nodes tend to be in the transmission range of each other, even after DoS, which, in turn, results in more number of edges in the graph after the removal of nodes that succumb to DoS, thereby increasing the probability that the graph is connected even under DoS attack. However, increase in p_{DoS} results in removal of more nodes and edges, and canstill disconnect the network with a high probability. For example, when $p_{\text{DoS}} > 0.85$, the network is almost surely disconnected for $\overline{p} \leq 0.06$.

Next, the effect of DoS on the network connectivity performance in the presence of defense mechanisms for DoS are studied, considering three defense mechanisms, (i) the individual detection mechanism proposed in [27], (ii) the centralized protocol developed in [27] and (iii) the distributed protocol developed in [28]. A network with 200 secondary users (i.e., $N = 200$ vertices in $G(N, \overline{p})$) is considered and the strength of DoS is increased by increasing the average number of malicious users, $E[Nm]$. For each value of $E[Nm]$, the probability of successful DoS, p_{DoS}, is computed for the individual detection mechanism, the centralized protocol (according to the analysis described in [27]) and for the distributed protocol in [28].

Figure 7.4 shows the probabilities that DoS disconnects an initially connected network, $p_{\text{disconnect}}$, after deploying DoS detection and mitigation

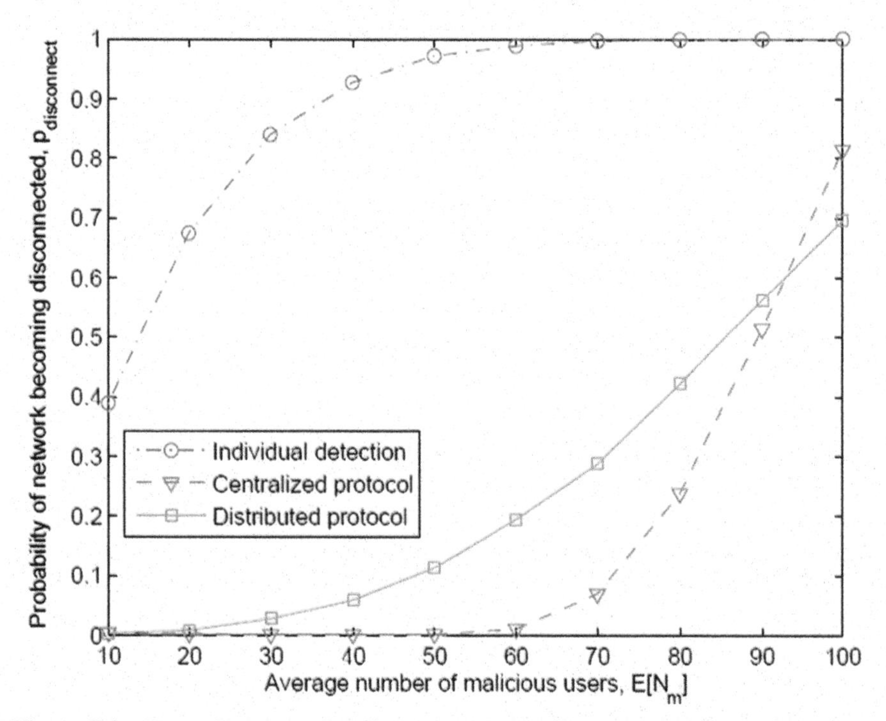

Figure 7.4 Comparison on probability that DoS disconnects an initially connected network after deploying detection and mitigation mechanisms. The probability that any two secondary users have a direct link is 0.0436. The legends "Individual detection" represents the individual detection mechanism described in [27]. "Centralized protocol" and "Distributed protocol" represent the centralized and the distributed protocols proposed in [27] and [28], respectively.

mechanisms[2]. It is observed from Figure 7.4 that different DoS detection and mitigation mechanisms exhibit significantly different impact on keeping the network connected. While the network implementing only the individual detection mechanism shows some resilience to DoS compared to the network with no defense mechanisms against DoS[3], its ability to keep the network connected under DoS is still poor. For example, even with 20 malicious users launching DoS, the probability of network becoming disconnected is about 0.7. This shows that when network level impact is considered, it makes more sense for a collaborative approach to spectrum decision than taking an individual decision. The results shown in Figure 7.4 indicate that the centralized protocol proposed in [27] and the distributed protocol proposed in [28] are effective in maintaining the network connected (and hence, increase reliability) under DoS. For example, when the network has no more than 60 malicious users, the centralized protocol can almost surely maintain the network connected under DoS. When the number of malicious users increases, the centralized protocol can still reduce the probability of the network becoming disconnected due to PUEA by about 20%. The distributed protocol proposed in [28] can improve the probability of the network becoming disconnected due to DoS by about 31% when the number of malicious users is large. For small number of malicious users, the distributed protocol can improve the probability that DoS disconnects the network, by two orders of magnitude.

It is observed that the performance of the distributed protocol is better than that of the centralized protocol for lower loads of malicious users. To understand this behavior in more detail, the probability of successful DoS when deploying the distributed protocol proposed in [28] with that when deploying the centralized protocol proposed in [27] are compared and depicted in Figure 7.5. It is observed that the distributed protocol outperforms the centralized protocol in terms of successfully detecting DoS when the expected number of malicious users $E[Nm] < 45$. For larger number of malicious users, the centralized protocol results in smaller probability of successful DoS and therefore, smaller probability of a disconnected ad hoc network under DoS. The reason for this behavior is explained below.

In a network with N nodes, let the probability that any node succumbs to DoS while implementing the individual detection mechanism alone, be p_{DoS}. While deploying the centralized protocol in [27], a node finally succumbs

[2]The legends "Individual detection" represents the individual detection mechanism described in [27], while "Centralized protocol" and "Distributed protocol" represent the centralized and the distributed protocols proposed in [27] and in [28], respectively.

[3]In such a network, every time DoS is launched, it will succeed with probability one.

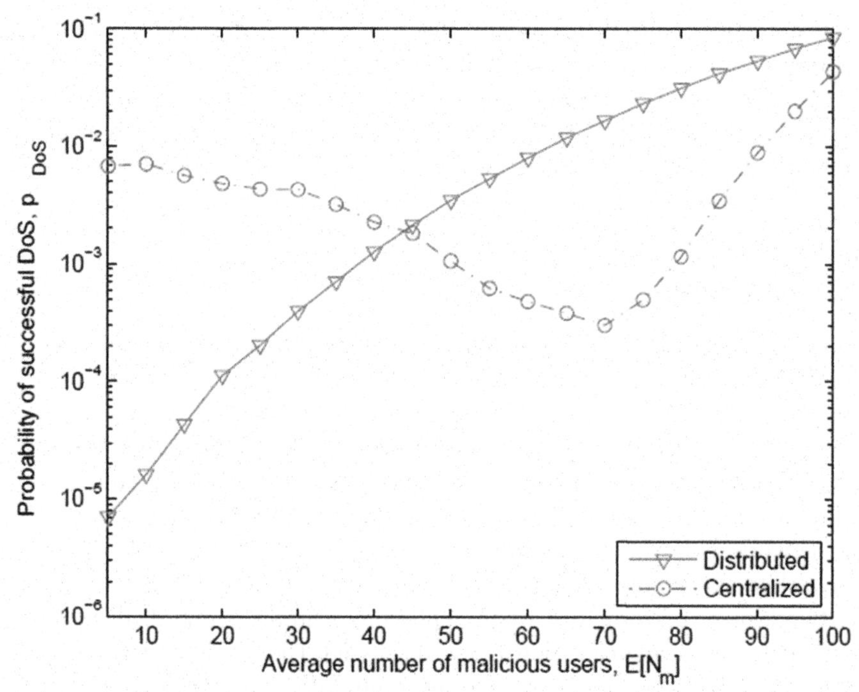

Figure 7.5 Probability of successful DoS by using the centralized protocol in [27] and the distributed protocol presented in [28].

to DoS if the fraction of nodes individually succumbing to DoS (using the individual detection mechanism alone) is greater than a specified threshold, i.e., the probability of successful DoS while deploying the centralized protocol, $p_{\text{DoS}}^{\text{centralized}}$, is approximately written as

$$p_{\text{DoS}}^{\text{centralized}} \approx \mathrm{P_r}\left\{N_{p_{\text{DoS}}} > thershold\right\} \tag{7.9}$$

In a network deploying the distributed protocol presented in [28], a node succumbs to DoS if the node itself and ALL its neighbors succumb to DoS while making the individual decision. Therefore, the probability of successful DoS while deploying the distributed protocol with a node with N_0 neighbors, $p_{\text{DoS}}^{\text{distributed}}$, can be written as

$$p_{\text{DoS}}^{\text{distributed}} = \left(p_{\text{DoS}}\right)^{N_0+1} \tag{7.10}$$

When the number of malicious users is small, each node can detect DoS better while implementing the individual detection mechanism alone, i.e., p_{DoS} is small. For smaller values of p_{DoS}, the value of $p_{\mathrm{DoS}}^{\mathrm{distributed}}$ in Equtaion (7.10) is negligible and is likely to be smaller than the value of $p_{\mathrm{DoS}}^{\mathrm{centralized}}$ in Equation (7.9). However, when the number of malicious users increase, p_{DoS} increases and the value of $p_{\mathrm{DoS}}^{\mathrm{distributed}}$ in Equation (7.10) maybe significantly larger than that of $p_{\mathrm{DoS}}^{\mathrm{centralized}}$ in Equation (7.9), resulting in better performance for the centralized protocol.

7.5 Conclusions

This chapter analyzed the network layer performance of ad hoc 5G wireless networks under denial of service attacks, in terms connectivity. The ad hoc network was modeled as a random geometric graph and an approximate Erdös-Rényi graph based analysis was performed to determine the probability that DoS disconnects a connected ad hoc network. Numerical results indicated that although connectivity of the network under DoS can be enhanced in a denser network, DoS should always be treated with defense mechanisms for the network to stay connected when nodes are more vulnerable to the attack. The network layer performance when deploying centralized and distributed defense mechanisms against DoS was also discussed. The distributed protocol reduced the probability of the network becoming disconnected by 31% to two orders of magnitude. The centralized protocol was shown to prevent the network from becoming disconnected when the number of malicious users is small. When the number of malicious users increases, the centralized protocol can reduce the probability of DoS disconnecting the network, by about 20%.

References

[1] UMTS, "Mobile traffic forecasts 2010–2020," Report, UMTS Forum, Jan. 2011.

[2] D. R. C. S. E. C. Limited, "5G Vision," White Paper, Feb. 2015. [Online]. Available: http://www.samsung.com/global/business-images/insights/2015/Samsung-5G-Vision-0.pdf

[3] H. Technologies, "U-LTE: Unlicensed spectrum utilization of LTE." [Online]. Available: http://www.huawei.com/ilink/en/download/HW\327803

[4] B. Munson, "Washington glowing over AWS-3 auction results," Wireless Week, Jan. 2015.

[5] "Cisco visual networking index: Global mobile data traffic forecast update 2014 2019," Cisco White Paper. [Online]. Available: http://cisco.com/c/en/us/solutions/collateral/service-provider/visual-networking-indexvni/whitepaperc11-520862.html

[6] C. Cordeiro, K. Challapali, and M. Ghosh, "Cognitive phy and mac layers for dynamic spectrum access and sharing of tv bands," Proceeding, First Intl. Workshop on Technol. and Policy for Accessing Spectrum (TAPAS) 2006, Aug. 2006.

[7] R. Chen, J. M. Park, and K. Bian, "Robust distributed spectrum sensing in cognitive radio networks," Proceedings, IEEE Conference on Computer Communications (INFOCOM'2008), pp. 1876–1884, Apr. 2008.

[8] G. Fodor, "D2D communications: What part will it play in 5G," Jul. 2014. [Online]. Available: http://www.ericsson.com/research-blog/5g/device-device-communications/

[9] X. Lin, J. G. Andrews, A. Ghosh, and R. Ratasuk, "An overview of 3GPP device-to-device proximity services," IEEE Commun. Mag., vol. 52, no. 4, pp. 40–48, Apr. 2014.

[10] Q. Ye, M. Al-Shalash, C. Caramanis, and J. G. Andrews, "Resource optimization in device-to-device cellular systems using time-frequency hopping." [Online]. Available: http://arxiv.org/abs/1309.4062

[11] X. Lin, J. G. Andrews, and A. Ghosh, "Spectrum sharing for device-to-device communications in cellular networks." [Online]. Available: http://arxiv.org/abs/1305.4219

[12] "IEEE Standards for information technology – Telecommunications and information exchange between systems – Wireless Regional Area Networks-Specific Requirements – Part 22-Cognitive wireless RAN medium access control (MAC) and physical layer (PHY) specifications: Policies and procedures for operation in the TV bands," Jun. 2006.

[13] Al-Dulaimi, S.l-Rubaye, Q. Ni, and E. Sousa, "5G communications race: Pursuit of more capacity triggers LTE in unlicensed band," IEEE Vehic. Technol. Mag., vol. 10, no. 1, pp. 43–51, Feb. 2015.

[14] M. Hangargi, "Business need for security: Denial of service attacks in wireless networks," Masters Dissertation, Department of Information Assurance, North Eastern University, Apr. 2012.

[15] Y. Li, B. Kaur, and B. Andersen, "Denial of service prevention for 5G," Wireless Personal Commun., vol. 57, no. 3, pp. 365–376, Apr. 2011.

[16] M. Klassen and N. Yang, "Anomaly based intrusion detection in wireless networks using Bayesian classifier," Proc., Intl. Conf. on Advanced Computational Intelligence (ICACI'2012), Sept. 2012.

[17] S. Anand, S. Sengupta, K. Hong, K. P. Subbalakshmi, R. Chandramouli, and H. Cam, "Exploiting channel fragmentation and aggregation/bonding to create security vulnerabilities," IEEE Trans. on Vehic. Technol., vol. 63, no. 8, pp. 3867–3874, Oct. 2014.
[18] R. Chen and J. M. Park, "Ensuring trustworthy spectrum sensing in cognitive radio networks," Proceedings, IEEE Workshop on Networking Technologies for Software Defined Radio Networks, pp. 110–119, Sep. 2006.
[19] R. Chen, J. M. Park, and J. H. Reed, "Defense against primary user emulation attacks in cognitive radio networks," IEEE Journal on Selected Areas in Communications: Special Issue on Cognitive Radio Theory and Applications, vol. 26, no. 1, pp. 25–37, Jan. 2008.
[20] M. Wyglinski, M. Nekovee, and Y. T. Hou, Cognitive Radio Communications and Networks: Principles and Practice. Elsevier Inc., 2010.
[21] S. Anand, Z. Jin, and K. P. Subbalakshmi, "An analytical model for primary user emulation attacks in cognitive radio networks," Proceedings, IEEE Symposium of New Frontiers in Dynamic Spectrum Access Networks (DySPAN'2008), Oct. 2008.
[22] Z. Jin, S. Anand, and K. P. Subbalakshmi, "Detecting primary user emulation attacks in dynamic spectrum access networks," Proceedings, IEEE International Conference on Communications (ICC'2009), Jun. 2009.
[23] ——, "Mitigating primary user emulation attacks in dynamic spectrum access networks using hypothesis testing," ACM SIGMOBILE Mobile Computing and Communications Review, Special Issue on Cognitive Radio Technologies and Systems, vol. 13, no. 2, pp. 74–85, April 2009.
[24] ——, "Performance analysis of dynamic spectrum access networks under primary user emulation attacks," Proceedings, IEEE Global Communications Conference (GLOBECOM'2010), Dec. 2010.
[25] ——, "Impact of primary user emulation attacks on dynamic spectrum access networks," IEEE Trans. on Commun., vol. 60, no. 9, pp. 2635–2643, Sep. 2012.
[26] F. Xing and W. Wang, "Understanding dynamic denial of service attacks in mobile ad hoc networks," Proc., IEEE Military Commun. Conf. (MILCOM'2006), Oct. 2006.
[27] Z. Jin, S. Anand, and K. P. Subbalakshmi, "Robust spectrum decision protocol against primary user emulation attacks in dynamic spectrum access networks," Proceedings, IEEE Global Communications Conference (GLOBECOM'2010), Dec. 2010.

[28] ——, “NEAT: A NEighbor AssisTed spectrum decision protocol for resilience against PUEA,” Cognitive Radio Eletromagnetic Spectrum Security (CRESS’2014), Oct. 2014.

[29] M. Penrose, Random Geometric Graphs. Oxford University Press, 2003.

[30] B. Bollobás, Random Graphs, 2nd ed. Cambridge University Press, 2001.

[31] D. Lu, X. Huang, P. Li, and J. Fan, “Connectivity of large-scale cognitive radio ad hoc networks,” Proc., IEEE Intl. Conf. on Comp. Commun. (INFOCOM’2012), Mar. 2012.

About the Authors

Zituo Jin (SM'08) is currently pursuing his Master's program in financial management at the Courant Institute of Mathematical Sciences in New York University (NYU). He graduated with a Ph.D. from the Department of Electrical and Computer Engineering at Stevens Institute of Technology, Hoboken, New Jersey. He received the B.E. degree in the Department of Information Engineering from Xi'an Jiaotong University, Xi'an, China, in 2006.

His current research interests include cognitive radio network security and denial-of-service attack in dynamic spectrum access networks. He is a student member of IEEE, IEEE communications society and IEEE computer society.

Santhanakrishnan Anand received his Ph.D. degree from the Indian Institute of Science, Bangalore, India, in 2003. He is currently an Assistant Professor at Department of Electrical and Computer Engineering, New York Institute of Technology.

His current areas of research include spectrum management and security in next generation wireless networks, covert timing channels, social media analytics, dynamics of Wikipedia and information propagation in Internet media.

Dr. Santhanakrishnan Anand received the Seshagiri Kaikini medal for the best Ph.D. dissertation in the electrical sciences division, Indian Institute of Science for the academic year 2003–2004. He has represented Samsung Electronics in 3GPP SA2 and IEEE 802.20 standardization meetings.

Koduvayur P. Subbalakshmi (Suba) is a Professor at Stevens Institute of Technology. She will serve as a Jefferson Science Fellow at the US Department of State during the academic year 2016–2017.

Her research interests include: Cognitive radio networks, Cognitive Mobile Cloud Computing, Social Media Analytics and Wireless security. She is a Subject Matter Expert for the National Spectrum Consortium.

She is a Founding Associate Editor of the IEEE Transactions on Cognitive Communications and Networking and the Founding Chair of the Security Special Interest Group of the IEEE Technical Committee on Cognitive Networks. She is also a recipient of the NJIHOF Innovator award.

Rajarathnam Chandramouli (Mouli) is the Thomas Hattrick Chair Professor of Information Systems in Electrical and Computer Engineering (ECE) and a Professor in the School of Systems and Enterprises at Stevens Institute of Technology. He is the Founding Director of NSF SAVI: Institute for Cognitive Networking and the Co-Director of the Information Networks and Security (iNFINITY) laboratory. Prior to joining Stevens he was on the ECE faculty at Iowa State University, Ames.

His research covers cognitive radio networking, dynamic spectrum management/access, text analytics and forensics, social media analytics and security, and prototyping/experimental research in these areas.

His research and technology commercialization projects are funded by the National Science Foundation, National Institute of Justice, Department of Defense and the industry.

8

Optimal Signal Design for Wavelet Radio TOA Locationing with Synchronization Error for 5G Networks

Homayoun Nikookar

Netherlands Defence Academy, the Netherlands

8.1 Introduction

High throughput and low cost and latency as well as high reliability, spectral efficiency and flexibility are, among others, major requirements of 5G. Another remarkable issue for the 5G networks is its context-awareness. Context can be location. Location-based services are becoming more and more important. In this chapter accurate positioning of devices with 5G using flexible wavelet packet modulation is addressed. Wavelet Packet modulated (WPM) signal can be used for joint communications and ranging (two functionalities in one technology). Using wavelet technology the optimal signal will be designed for the time-of-arrival (TOA) or time-difference-of-arrival (TDOA) locationing when there exists a synchronization error. The focus will be on the wavelet packet modulated signal as wavelets have lower sensitivity to distortion and interference as a result of synchronization error. Further advantage of wavelet technology lies in its flexibility to customize and shape the characteristics of the waveforms for 5G radio communication and ranging purposes. In addition to the flexibility, the reconfigurability of the wavelet signals is also an important characteristic for the successful application of this technology in cognitive radio communication and ranging systems.

There are basically three methods to estimate the position with radio signals: Received Signal Strength (RSS), Angle of Arrival (AOA) and Time of Arrival (TOA) [1, 2]. Here we focus on the TOA method. In this location technique the time of arrival of the signal sent by the mobile agent to be positioned is measured at each receiver (access point). The propagation time of each signal is known and is proportional to the distance. As shown in Figure 8.1 the measured time provides information in a set of points around the circumference of a circle having the radius of distance between the object (mobile) and the access point. The intersection of the circles is the mobile's position. Similar to the RSS method, for the 2-D location based on the TOA technique, at least three access points are required. Let t_1, t_2 and t_3 denote the flight time of the transmitted signal of the object (mobile) to be positioned, to the respective receivers (access points). The access points 1, 2 and 3 are respectively positioned at locations (0,0), $(0,y_2)$ and (x_3,y_3). Let (x,y) denote the coordinates of the handset (object) to be positioned. By estimating the time of arrivals, the following set of equations should be solved to obtain the position (x,y) of the mobile devices.

$$\begin{aligned} d_1 &= ct_1 = \sqrt{x^2 + y^2} \\ d_2 &= ct_2 = \sqrt{x^2 + (y - y_2)^2} \\ d_3 &= ct_3 = \sqrt{(x - x_3)^2 + (y - y_3)^2} \end{aligned} \tag{8.1}$$

where c is the speed of light. It should be mentioned that to avoid ambiguity in identifying the intersection of circles in Figure 8.1 all three equations in (8.1) must be considered.

The key factor in time based distance estimator is the arrival time of the first path. TOA utilizes the time delay to get the distance and looks for the intersection of at least three circles to estimate the location. It requires synchronization at both transmitter and receiver side, which is always an important issue in the wireless network, as well as the information of transmission time. If the system is not synchronized or if there is an offset in the time of transmission, the TOA methods cannot work properly. For instance, even a 1msec inaccuracy in the TOA estimation can cause an error of up to 300 m in ranging!

A variant of TOA, the time difference of arrival (TDOA) scheme, can improve the situation. It calculates the target mobile's position according to the time differences between each measurements, rather than the time measurement itself as in TOA. Therefore, as shown in Figure 8.2, the TDOA searches the hyperbolic intersection and only needs the receiver clock synchronization without the information of time of transmission. The equations for the TDOA positioning become:

$$\begin{aligned} &\sqrt{(x - x_i)^2 + (y - y_i)^2} - \sqrt{(x - x_j)^2 + (y - y_j)^2} = d_i - d_j, \\ &i = 1, 2, 3 \quad \text{and} \quad j = 1, 2, 3 \end{aligned} \tag{8.2}$$

Efficient methods to solve these nonlinear equations have been reported [3].

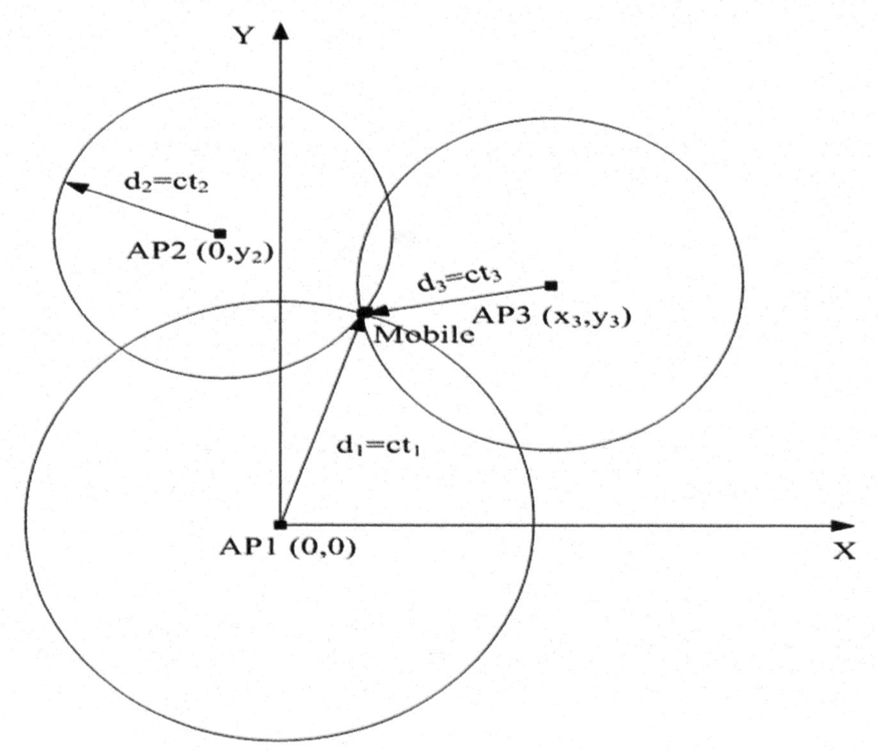

Figure 8.1 The TOA positioning method (APi: *i*th Access Point).

The RSS method is not an accurate method in locationing, especially in multipath fading environments. The AOA method requires a complex hardware. The TOA and TDOA methods are more appropriate for the positioning (especially when the bandwidth of the transmission is wide providing a fine time resolution). A remarkable issue in locationing with the TOA or TDOA techniques is the estimation of the arrival time of the transmitted signals. Major TOA estimation techniques use the correlation function and find its maximum to estimate the TOA or TDOA [4].

In this chapter we shall design the optimal signal for TOA or TDOA locationing when there exists a synchronization error. We particularly concentrate on the wavelet packet modulated signal for communications and ranging[1] as wavelets have lower sensitivity to distortion and interference (due to

[1]It should be emphasized that the Wavelet Packet Modulated (WPM) signal can be used for joint communications and ranging. This joint functionality in one technology is the outstanding capability of the WPM signal.

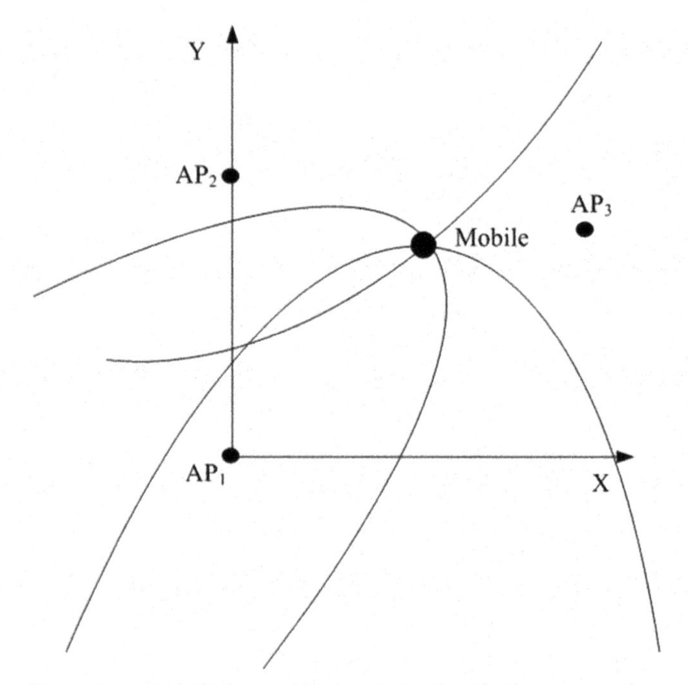

Figure 8.2 The TDOA positioning method (APi: *i*th Access Point).

synchronization error) and also provide flexibility in signal design when there exists a synchronization error in the received signal. The advantage of wavelet transform lies in its flexibility to customize and shape the characteristics of the waveforms for radio communication and ranging purposes. In addition to the flexibility, the reconfigurability of the wavelet signals is also an important characteristic of this technology which paves the way for its successful application in cognitive radio communication and ranging systems of future.

8.2 Wavelet Signal Design

This Section focuses on the design steps of wavelet signal. The implementation of filter also will be discussed in this section.

8.2.1 Design Procedure

The attributes of the wavelet packet modulation system greatly depend on the set of transmission bases utilized which in turn is determined by the filters used. This means that by adapting the filters one can adapt the transmit waveform characteristics to satisfy a system specification. Choosing the right filter though is a delicate task. The filters cannot be arbitrarily chosen and instead have to satisfy a number of constraints. Besides the design objectives there are other budgets which have to be considered in order to guarantee that the designed wavelet is valid. The design procedure consists of 3 major steps [5], namely:

1. Formulation of design problem, i.e., stating the design objectives and constraints mandated by wavelet theory.
2. Application of suitable optimizations and transformations to make the problem tractable.
3. Utilization of numerical solvers to obtain the required filter coefficients.

At the end of the design procedure a low pass Finite-Impulse-Response (FIR) filter $h[n]$, satisfying the design and wavelet constraints, is obtained. From this filter the other three filters $g[n]$, i.e., the high-pass filter, $h'[n]$, i.e., the dual of $h[n]$ and $g'[n]$, i.e., the dual of $g[n]$ are derived through the Quadrature Mirror Filter (QMF) relation [5], i.e., $g[n] = (-1)^n h[L-1-n]$, for $h[n]$ of length L. Where $h'[n] = h^*[-n]$, and $g'[n] = g^*[-n]$.

In the following sections we will elaborate on each of these processes.

8.2.2 Filter Bank Implementation of Wavelet Packets

It is well known that compactly supported orthonormal wavelets can be obtained from a tree structure constructed by successively iterating discrete two-channel paraunitary filter banks [5–8]. Time and frequency limited orthonormal wavelet packet bases $\xi(t)$ can be derived by recursively iterating discrete half-band high $g[n]$ and low-pass $h[n]$ filters, as[2]:

$$\begin{aligned} \xi_{l+1}^{2p}(t) &= \sqrt{2}\sum_m h\left[m\right]\xi_l^p\left(2t-m\right) \\ \xi_{l+1}^{2p+1}(t) &= \sqrt{2}\sum_m g\left[m\right]\xi_l^p\left(2t-m\right) \end{aligned} \tag{8.3}$$

In (8.3) the subscript l denotes the level in the wavelet tree structure and superscript p indicates the waveform index.The number of bases p generated is determined by the number of iterations l of the two-channel filter bank. Equation (8.3), known as 2-scale equation, can be interpreted as follows – a basis function belonging to a certain subspace of lower resolution can be obtained from shifted versions of the bases belonging to a subspace of higher resolution; and the weights h and g used in the transformation are low and high pass in nature. The filters h and g form a quadrature mirror pair and are also known as analysis filters. These filters have duals/adjoints known

[2]The expressions are considered in continuous time domain to convenience derivations in Section 8.4.

as synthesis filters which are also a pair of half-band low h' and high pass filters g'. All these four filters share a strict and tight relation and hence it is enough if the specifications of one of these filters are available. The wavelet packet sub-carriers (used at the radio transmitter end) are generated from the synthesis filters. And the wavelet packet duals (used at the radio receiver end) are obtained from the analysis filters. The entire wavelet packet modulated (WPM) transceiver structure can thus be realized by this set of two QMF pairs. Hence, the design process can also be confined to the construction of one of the filters, usually the low pass analysis filter *h*.

8.3 Problem Statement

The time synchronization error in the received signal for the data detection of communications and the TOA or TDOA estimation of ranging is modeled by shifting the received data samples R[n] by a time offset Δ_t to the left or right as:

$$R[n \pm \Delta_t] = S[n] + w[n]. \tag{8.4}$$

Here, S[n] denotes the transmitted signal and w[n] the additive white Gaussian noise (AWGN). According to wavelet theory [5] under ideal conditions, when the WPM transmitter and receiver are perfectly synchronized and the channel is benign, the detection of signal in the uth symbol and kth sub-carrier $\hat{a}_{u',k'}$ is the same as the transmitted data $a_{u,k}$[3]. However, errors are introduced in the demodulation decision making process under time offset errors Δ_t as elucidated below:

$$\begin{aligned}
\hat{a}_{u',k'} &= \sum_{n} R[n]\xi_l^{k'}[(u'N - n + \Delta_t] \\
&= \sum_{n}\sum_{u}\sum_{k=0}^{N-1} a_{u,k}\xi_l^{k}[n - uN]\xi_l^{k'}[u'N - n + \Delta_t] \\
&= \sum_{u}\sum_{k=0}^{N-1} a_{u,k}\left(\sum_{n}\xi_l^{k}[n - uN]\xi_l^{k'}[u'N - n + \Delta_t]\right)
\end{aligned} \tag{8.5}$$

Where N is the number of wavelet subcarriers. Defining the cross waveform function $\Omega(\Delta_t)$ as:

$$\Omega_{k,k'}^{u,u'}[\Delta_t] = \sum_{n}\xi_l^{k}[n - uN]\xi_l^{k'}[u'N - n + \Delta_t] \tag{8.6}$$

[3]The apostrophes in the symbol u' and carrier k' indices are used to indicate receiver side.

the demodulated data corrupted by the interference due to loss of orthogonality at the receiver for the kth subcarrier and uth symbol can be expressed as:

$$\hat{a}_{u',k'} = \underbrace{a_{u',k'}\Omega^{u',u'}_{k',k'}[\Delta_t]}_{\text{Desired Alphabet}} + \underbrace{\sum_{u;u\neq u'} a_{u,k'}\Omega^{u,u'}_{k',k'}[\Delta_t]}_{\text{ISI}}$$
$$+ \underbrace{\sum_{u}\sum_{k=0;k\neq k'}^{N-1} a_{u,k}\Omega^{u,u'}_{k,k'}[\Delta_t]}_{\text{IS-ICI}} + \underbrace{w_{u',k'}}_{\text{Gaussian Noise}} \quad (8.7)$$

For the data communication, in (8.7), the first term stands for the attenuated useful signal, the second term denotes Inter Symbol Interference (ISI), third term gives Inter Symbol Inter Carrier Interference (IS-ICI) and the last term stands for Gaussian noise. Generally speaking multi-carrier systems are highly sensitive to loss of time synchronization. A loss of time synchrony results in samples outside a WPM symbol getting selected erroneously, while useful samples at the beginning or at the end of the symbol getting discarded. It also introduces ISI and ICI causing performance degradation.

For the locationing, TOA or TDOA estimation methods use the correlation function and find its maximum to estimate the time of arrival of the signal to be used for positioning. However, because of time synchronization error there will be a cross correlation energy between the wavelets. Therefore, our objective in this paper is to design proper wavelets for TOA ranging when there is a time synchronization error. The design is achieved by minimizing the interference caused by the timing error.

We also note that though WPM and Orthogonal Frequency Division Multiplexing (OFDM) share many similarities as orthogonal multicarrier systems, they are significantly different in their responses to loss of time synchronization. This difference is a result of the fact that the WPM symbols overlap with each other and are longer than the OFDM symbol[4]. Under a loss in time synchronization, the overlap of the symbols in WPM causes each symbol to interfere with several other symbols while in OFDM each symbol interferes only with its neighbors. The second important difference is in the usage of guard intervals. OFDM benefits from the cyclic prefix which significantly improves its performance under timing errors. WPM cannot use guard intervals because of the symbol overlap.

[4]The length of the symbol and the degree of overlap is determined by the length of wavelet filter used.

Fortunately, WPM offers the possibility in adjusting the properties of the waveforms in a way that the errors due to loss of synchronization can be minimized. In Section 8.5 we present a method to design a new family of wavelet filters which minimize the energy of the timing error interference for ranging applications. But before that, let us discuss the major wavelet properties that are important in the selection of wavelets.

8.4 Important Wavelet Properties

Generally speaking the wavelet tool is a double edged sword – on one hand there is scope for customization and adaptation; on the other hand there are no clear guidelines to choose the best wavelet from for a given application. In order to ease the selection process constraints, such as orthogonality, compact support and smoothness are imposed. Here we shall discuss them in more detail.

8.4.1 Wavelet Existence and Compact Support

This constraint is necessary to ensure that the wavelet has finite non-zero coefficient and thus the impulse response of the wavelet decomposition filter is finite as well. According to [9], this property can be derived by simply integrating both sides of the two-scale Equation in (8.8) and can be derived as follows[5]:

$$\begin{aligned}\int_{-\infty}^{\infty}\xi(t)dt &= \sqrt{2}\int_{-\infty}^{\infty}\sum_{n}h[n]\xi(2t-n)dt\\ \int_{-\infty}^{\infty}\xi(t)dt &= \sqrt{2}\sum_{n}h\left[n\right]\int_{-\infty}^{\infty}\xi(2t-n)dt\\ \int_{-\infty}^{\infty}\xi(t)dt &= \sqrt{2}\sum_{n}h\left[n\right]\int_{-\infty}^{\infty}0.5\xi(2t-n)d(2t-n)\end{aligned} \tag{8.8}$$

Substituting $u = 2t - n$, (8.8) can be rewritten as:

$$\begin{aligned}\int_{-\infty}^{\infty}\xi(t)dt &= \frac{1}{\sqrt{2}}\sum_{n}h\left[n\right]\int_{-\infty}^{\infty}\xi(u)du\\ \frac{\int_{-\infty}^{\infty}\xi(t)dt}{\int_{-\infty}^{\infty}\xi(u)du} &= \frac{1}{\sqrt{2}}\sum_{n}h\left[n\right]\end{aligned}$$

[5]The subscripts denoting the decomposition level l and the waveform index p have been dropped for convenience.

Finally we obtain the compactly supported wavelet constraint as:

$$\sum_{n} h[n] = \sqrt{2} \tag{8.9}$$

It should be noted that the derivation resulting in (8.9) is also recognized as the wavelet existence constraint.

8.4.2 Paraunitary Condition

The paraunitary or the orthogonality condition is essential for many reasons. First, it is a prerequisite for generating orthonormal wavelets [6, 7]. Second, it automatically ensures perfect reconstruction of the decomposed signal. The constraint can be derived using the orthonormality property of the scaling function and its shifted version as follows:

$$\int_{-\infty}^{\infty} \xi\,(t)\,\xi\,(t-k)\,dt = \delta[k] \tag{8.10}$$

Substituting the two-scale Equation (8.3) in (8.10) we get:

$$\int_{-\infty}^{\infty} \sum_{n} h\,[n]\,\xi(2t-n)\sqrt{2} \sum_{m} h\,[m]\,\xi(2(t-k)-m)\sqrt{2}dt = \delta[k]$$

$$2\sum_{n} h\,[n] \sum_{m} h\,[m] \int_{-\infty}^{\infty} \xi(2t-n)\xi(2(t-k)-m)dt = \delta[k]$$

$$2\sum_{n} h\,[n] \sum_{m} h\,[m] \int_{-\infty}^{\infty} 0.5\xi(2t-n)\xi(2(t-k)-m)d(2t) = \delta[k] \tag{8.11a}$$

Or

$$\sum_{n} h\,[n]\,h\,[n-2k] = \delta[k] \quad \text{for } k = 0, 1, \ldots, (L/2) - 1 \tag{8.11b}$$

Equation (8.11b) is called double shift orthogonality relation of the wavelet low pass filters impulse responses. In (8.11b), L illustrates the length of the low pass wavelet filter impulse response. For a filter of length L the orthogonality condition (8.11b) imposes $^{L}/_{2}$ non-linear constraints on $h[n]$.

8.4.3 Flatness/*K*-Regularity

This property is a rough measure of smoothness of the wavelet. The regularity condition is needed to ensure that the wavelet is smooth in both time and frequency domains [10]. It is normally quantified by the number of times a wavelet is continuously differentiable. The simplest regularity condition is the *flatness* constraint which is stated on the low pass filter. A low pass filter (LPF), $h[n]$, is said to satisfy Kth order flatness if its transfer function $H(\omega)$ contains K zeroes located at the Nyquist frequency ($\omega = \pi$). For any function $Q(\omega)$ with no poles or zeros at ($\omega = \pi$) this can be written as:

$$H(\omega) = \left(\frac{1 + e^{j\omega}}{2}\right)^K Q(\omega) \quad \text{with } Q(\pi) \neq 0 \tag{8.12}$$

In (8.12), $Q(\omega)$ is a factor of $H(\omega)$ that does not have any single zero at $\omega = \pi$. Having K number of zeros at $\omega = \pi$ also means that $H(\omega)$is K-times differentiable and its derivatives are zero when they are evaluated at $\omega = \pi$. Considering that:

$$H(\omega) = \sum_n h[n] \exp(-j\omega n), \tag{8.13}$$

the kth order derivative of $H(\omega)$ would be:

$$H^{(k)}(\omega) = \sum_n h[n] (-jn)^k \exp(-j\omega n) \tag{8.14}$$

The evaluation of (8.14) at $\omega = \pi$ would result in:

$$H^{(k)}(\pi) = \sum_n h[n] (-jn)^k \exp(-j\pi n)$$

$$\sum_n h[n] (-j)^k (n)^k (e^{-j\pi})^n = 0$$

$$\sum_n h[n] (-1)^n (n)^k = 0$$

Therefore, the K-regularity constraint in terms of the low pass filter coefficients can be given as:

$$\sum_n h[n] (n)^k (-1)^n = 0 \quad \text{for } k = 0, 1, 2, \ldots, K - 1 \tag{8.15}$$

8.4.4 Degrees of Freedom to Design

The criterion (8.9), (8.11b) and (8.15) are necessary and sufficient conditions for the set to form an orthonormal basis and these conditions have to be

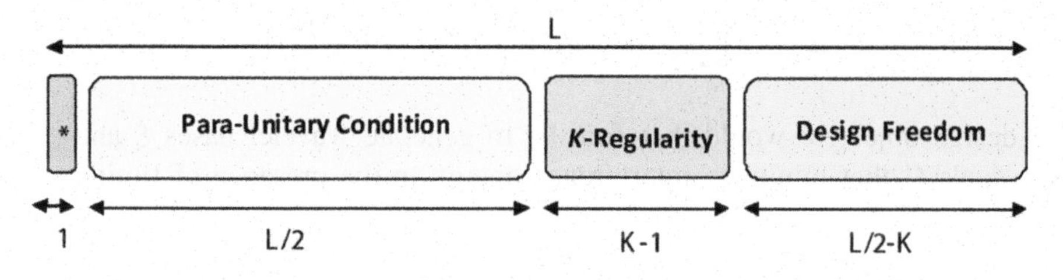

Figure 8.3 Wavelet conditions and degrees of freedom for design.

imposed for all design procedures. For a filter of length L this is essentially getting L unknown filter variables from L equations. Of these L equations, one equation is required to satisfy the wavelet existence condition, $^{L}/_{2}$ come from the paraunitaryness constraint, $K - 1$ from the regularity constraint and the remaining $^{L}/_{2}$-K conditions offer the possibility for establishing the design objective. The larger the value of $^{L}/_{2}$-K, the greater the degree of freedom for design and the greater is the loss in regularity. There is therefore a trade-off on offer. The $^{L}/_{2}$-K degrees of freedom that remain after satisfying wavelet existence, orthogonality and K-regularity condition can be used to design a scaling filter with the desired property (refer to Figure 8.3). In the next section we illustrate this with a special signal design example for the ranging application of wavelet packet modulation signal.

8.5 Formulation of Design Problem

In this Section, different aspects for formulation of design problem are discussed. Details on how the mathematical constraints converted from non-convex problem to the convex-problem are presented in this section as well.

8.5.1 Design Criterion

In disturbance-free environments the cross-correlations of WPM waveforms equal zero and perfect reconstruction is possible despite the time and frequency overlap. The timing error Δ_t on the other hand leads to the loss of the orthogonality between the waveforms and consequently they begin to interfere one with another leading to ICI and ISI error for communication and degrading the TOA estimation performance for ranging, stated as:

$$\Omega_{k,k';k\neq k'}^{u,u'}[\Delta_t] = \sum_n \xi_l^k(n - uN)\xi_l^{k'}(u'N - n + \Delta_t) \tag{8.16}$$

The design objective would therefore be to generate wavelet bases ξ and their duals ξ' that minimize interference energy in the presence of timing error:

$$\text{MINIMIZE:} \sum_{u,k;k\neq k'} \left|\Omega_{k,k'}^{u,u'}[\Delta_t]\right|^2 \text{ with respect to } \{\xi, \xi'\} \tag{8.17}$$

8.5.2 Wavelet Domain to Filter Bank Domain

The waveforms are created by the multilayered tree structure filter bank. Using Parseval's theorem of energy conservation it can be easily proven that the total energy at each level is equal regardless of the tree's depth. Therefore, minimizing the interfering energy at the roots of the tree will automatically lead to the decrease of total interfering energy at the higher tree branches. Furthermore, the two-channel filter banks through the 2-scale equation are related, albeit explicitly, to the wavelet waveforms. Therefore, the design process can be converted into a tractable filter design problem. We should hence be able to minimize deleterious effects of time synchronization errors in wavelet radio ranging by minimizing the following cross-correlation function:

$$\sum_{\Delta_t} |r_{hg}[\Delta_t]|^2 = \sum_n |h[n]g[n - \Delta_t]|^2 = \sum_n |h[n]((-1)^n h[L - n + \Delta_t])|^2 \tag{8.18}$$

The design problem of minimizing the interference energy due to timing offset can now be formally stated as an optimization problem with objective function (8.18) and constraints (8.9), (8.11b) and (8.15), i.e.,

$$\begin{aligned}
\text{MINIMIZE:} & \sum_{\Delta_t} |r_{hg}[\Delta_t]|^2 \quad \text{with respect to } h[n] \\
& \sum_n h[n] = \sqrt{2} \\
\text{SUBJECTTO:} & \sum_n h[n]\,h[n - 2k] = \delta[k] \quad \text{for } k = 0, 1, \ldots, (L/2) - 1 \\
& \sum_n h[n]\,(n)^k(-1)^n = 0 \quad \text{for } k = 0, 1, 2, \ldots, K - 1
\end{aligned} \tag{8.19}$$

As obvious, the second constraint in (8.19) is non-linear and therefore the optimization problem becomes a non-convex one. Thus, the optimization problem as given above can only be solved by general purpose solvers. However, such solvers are susceptible to being trapped in local minima. In order to overcome this difficulty, some authors have suggested multiple starting point techniques or branch-and-bound method [11]. Moreover, general purpose algorithms cannot guarantee that the found result is a global minimum and furthermore when number of constraints increases these algorithms often fail to provide a valid solution. The objective function and constraints can be solved much more efficiently using convex optimization and semi-definite programs [12–18]. In the following sections we attempt to express the design constraints in convex form so that convex optimization tools can be employed to obtain the solution [19–21]. Similar to the wavelet signal design for the distributed radio sensing networks [22], we shall move to the auto-correlation domain ($r_h[k] = \sum_{m \in z} h[m]h[m+k]$) to simplify the minimization problem.

8.5.3 Transformation of the Mathematical Constraints from Non-convex Problem to a Convex/linear One

Fortunately, it is possible to transform the non-convex/non-linear equations into a linear/convex problem by reformulating the constraints in terms of the autocorrelation sequence $r_h[k]$, [23–25]:

$$r_h[k] = \sum_{m \in z} h[m]\, h[m+k] \tag{8.20}$$

Taking into account the inherent symmetry of the autocorrelation sequence it can be defined more precisely as:

$$r_h[l] = \sum_{n=0}^{L-l-1} h[n]\, h[n+l] \quad \text{for } l \geq 0 \tag{8.21}$$

In (8.21), L is the length of the FIR filter and the autocorrelation function is symmetric about $l = 0$; i.e.:

$$r_h[-l] = r_h[l] \tag{8.22}$$

The three constraints (8.9), (8.11b), and (8.15) are derived in terms of $r_h[l]$ in the following sub-sections.

8.5.3.1 Compact support or admissibility constraint

The compact support constraint in (8.3) can be rewritten as:

$$\sum_{n=0}^{L-1} \sum_{l=-n}^{L-n-1} h[n]\, h[n+l] = 2 \tag{8.23}$$

Reversing the order of the summation and considering the fact that the impulse response of filter $h[n]$ has non-zero values only at $0 \leq n \leq L-1$, we obtain:

$$\sum_{l=-(L-1)}^{L-1} \sum_{n=0}^{L-l-1} h[n]\, h[n+l] = 2 \tag{8.24}$$

The compact support constraint in (8.9) can then be rewritten as:

$$\sum_{l=-(L-1)}^{L-1} r_h\,[l] = 2 \tag{8.25}$$

Taking into consideration the double shift orthonormality property (see Equation (8.11b)) and the fact that the autocorrelation sequence is symmetric, we can simplify (8.25) further as:

$$r_h\,[0] + 2\sum_{l=1}^{L-1} r_h\,[l] = 2 \quad \text{or}$$

$$\sum_{l=1}^{L-1} r_h\,[l] = \frac{1}{2} \tag{8.26}$$

Equation (8.26) is the compactly supported wavelet constraint stated in terms of the autocorrelation sequence $r_h[l]$.

8.5.3.2 Double shift orthogonality constraint

The double shift orthogonality constraint presented in (8.11b), can be expressed in terms of the autocorrelation sequence $r_h[l]$ as follows:

$$\sum_{m} h[m]\, h[m+2k] = r_h[2k] = \delta[k] \tag{8.27}$$

It should be noted that (8.27) is obtained by applying $n - 2k = m$ in Equation (8.11b). Hence the final double shift orthogonality constraint in terms of autocorrelation sequence $r_h[l]$ is:

$$r_h[2k] = \delta[k] = \begin{cases} 1, & \text{for } k = 0 \\ 0, & \text{otherwise} \end{cases} \quad \text{with } k = 0, 1, \ldots, \left\lfloor \frac{L-1}{2} \right\rfloor \tag{8.28}$$

where $\lfloor \cdot \rfloor$ denotes the floor function or the integer part of $(\cdot)$ which is the greatest integer that is less than or equal to $(\cdot)$. Again we make use of the symmetry property to simplify it. In contrast to (8.11b) which was non-convex, (8.28) consists of linear equalities and is also convex.

8.5.3.3 *K*-Regularity constraint

The regularity constraint can be reformulated in terms of autocorrelation sequence $r_h[l]$ by considering the square of the absolute value of Equation (8.12); i.e.:

$$|H(\omega)|^2 = \left(\frac{1+e^{-j\omega}}{2}\right)^K \left(\frac{1+e^{j\omega}}{2}\right)^K |Q(\omega)|^2 \tag{8.29}$$

Requiring the transfer function $H(\omega)$ to have K zeros at Nyquist frequency $(\omega = \pi)$ is equivalent to requiring $|H(\omega)|^2$ to have $2K$ zeros at $\omega = \pi$. Taking into account the fact that $|H(\omega)|^2$ is the Fourier transform of autocorrelation sequence of $r_h[l]$, evaluating the $2k$th order derivative of $|H(\omega)|^2$ and making use of the symmetry property of the autocorrelation sequence $r_h[l]$, it can be easily shown that

$$\sum_{l=1}^{L-1} (-1)^l \,(l)^{2k} r_h\,[l] = 0 \quad \text{for } k = 0, 1, \ldots, K-1 \tag{8.30}$$

Equation (8.30) states the regularity constraint in terms of autocorrelation sequence $r_h[l]$.

The admissibility, paraunitary and K-regularity conditions are readily available in the auto-correlation domain (Equations (8.26), (8.28) and (8.30), respectively). Therefore, we only have to derive the objective function. Now, we know that

$$r_h[n] = \begin{cases} \sum\limits_{m=0}^{L-n-1} h[m]\, h[m+n] & n \geq 0 \\ r_h(-n) & n < 0 \end{cases} \tag{8.31}$$

and

$$\begin{aligned} r_g[n] &= \sum_{m=0}^{L-n-1} g[m]\, g[m+n] \quad \text{for } n \geq 0 \\ &= \sum_{m=0}^{L-n-1} ((-1)^m h[L-m])((-1)^{m+n} h[L-(m+n)]) \\ &= (-1)^n r_h[n] \end{aligned} \tag{8.32}$$

Applying the corollary[6]: *The sum of squares of a cross-correlation between two functions equals the inner product of the autocorrelation sequences of these two functions,* and the double shift orthogonality property:

$$[2x] = \delta[x] = \begin{cases} 1, & \text{for } x = 0 \\ 0, & \text{otherwise} \end{cases} \quad \text{where } x = 0, 1, \ldots, \left\lfloor \frac{L-1}{2} \right\rfloor \tag{8.33}$$

The cross-correlation function $r_{hg}[n]$ can be rewritten in terms of $r_h[n]$ as follows:

$$\sum_{n=0}^{L-1} |r_{hg}[n]|^2 = \sum_{n=0}^{L-1} r_h[n]\, r_g[n]$$

$$= \sum_{n=0}^{L-1} r_h[n]\left((-1)^n r_h[n]\right) = \underbrace{\sum_{x=0}^{(\frac{L}{2}-1)} (r_h[2x+1])^2}_{\text{Odd numbered values}} - \underbrace{\sum_{x=0}^{(\frac{L}{2}-1)} (r_h[2x])^2}_{\text{Even numbered values}} \tag{8.34}$$

$$= \sum_{n=0}^{(\frac{L}{2}-1)} (r_h[2n+1])^2 - 1$$

The new optimization problem can thus be stated as:

Minimize $\sum_{n=0}^{(\frac{L}{2}-1)} (r_h[2n+1])^2$ *subject to the wavelet constraints* (8.26), (8.28) and (8.30), i.e.,

$$\text{MINIMIZE:} \sum_{n=0}^{(\frac{L}{2}-1)} (r_h[2n+1])^2$$

$$\sum_{l=1}^{L-1} r_h[l] = \frac{1}{2}$$

$$\text{SUBJECTTO:}\; r_h[2k] = \delta[k], \quad \text{for } k = 0, 1, \ldots, \left\lfloor \frac{L-1}{2} \right\rfloor$$

$$\sum_{l=1}^{L-1} (-1)^l l^{2k} r_h[l] = 0, \quad \text{for } k = 0, 1, \ldots, K-1$$

---

[6]Proved in the Appendix.

Since the optimization problem posed above is linear it is also convex. Therefore, any linear or convex optimization tool can be used to solve this problem. In this case, we choose SeDuMi [26] as generic Semi Definite Programming (SDP) solvers to solve the optimization problem. SeDuMi stands for *self-dual minimization* as it implements a self-dual embedding technique for optimization over self-dual homogeneous cones [26]. It comes as an additional Matlab® package and can be used for linear, quadratic and semidefinite programming. Normally it requires a problem to be described in a primal standard form but with modeling languages like YALMIP (short for Yet Another LMI Parser) the optimization problems can be directly expressed in a user-friendly higher level language [27]. Thus YALMIP allows the user to concentrate on the high-level modeling without having to worry about low-level details. We have developed a filter optimization program that incorporates most of the available optimization routines for Matlab® and which relies on YALMIP to translate the problem into the standard form. From the autocorrelation sequence, the algorithm derives filter coefficients with length L having minimum phase property[7]. At the end of the design process the filter coefficients of the analysis LPF will be generated. From the analysis the low-pass-filter (LPF) $h[n]$, the high-pass-filter (HPF) $g[n]$ and the synthesis filters, LPF $h'[n]$ and HPF $g'[n]$, can be obtained through the quadrature Mirror Filter (QMF) equations. And from these set of filters the wavelets and their duals can be derived using the 2-scale Equation (3) .

8.6 Results and Analysis

In this section we present a few results to demonstrate the design procedure. As already mentioned, the main variables of the design process are the length and regularity order of the filter.

8.6.1 Frequency and Impulse Response of Designed Filter

We set the length of the filter to 20 though it is possible to design longer or shorter filters. The order of regularity chosen is 5, which is a compromise between optimization space and wavelet regularity. The impulse response of the designed optimal filter is illustrated in Figure 8.4 and numerical values of filter coefficients are given in Table 8.1. Although the optimal filter is designed in the autocorrelation domain, the minimum-phase time domain coefficients obtained satisfy all constraints mandated by the design process.

[7]We chose filters having minimum phase because they guarantee stability.

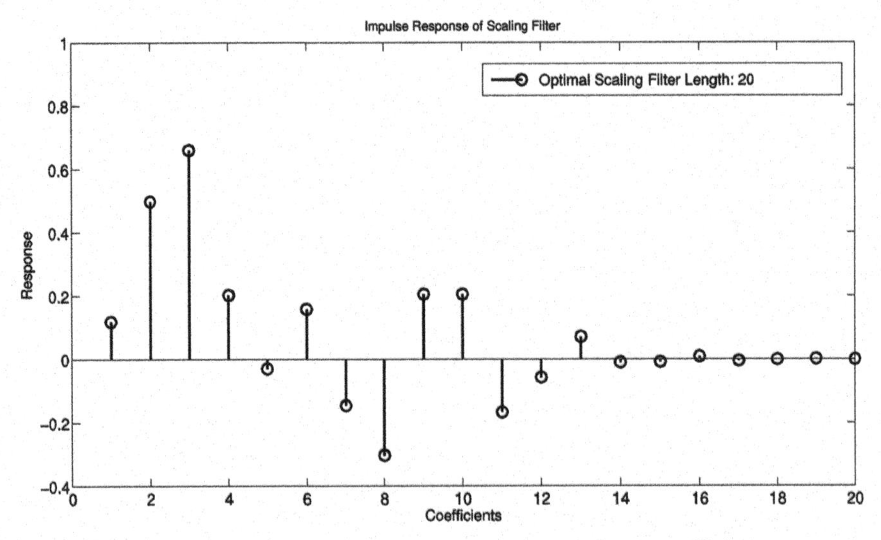

Figure 8.4 Impulse response of the optimal LPF with 20 coefficients.

Table 8.1 Optimal filter coefficients

0.119881851613898	0.498287367999060	0.660946808777660	0.203191803677134
–0.02915169068882	0.159448121842196	–0.144908809151642	–0.301681615791117
0.206305798368833	0.205999004857997	0.165410385138750	–0.0566148032177797
0.071282862607634	–0.00958254794419582	–0.0083940508446907	0.00912479119304040
–0.00498653232060	–0.000694408881953843	0.00154092796305564	–0.000370932610232259

The wavelet and scaling function of the newly designed optimal filter are illustrated in Figure 8.5a and 8.5b, respectively. The frequency response is shown in Figure 8.5c. Table 8.2 shows the specifications of the various filters used in the study along with the values of the corresponding objective functions. Clearly the designed wavelet has the lowest interference energy (due to time synchronization error).

8.6.2 Evaluation of Designed Filter under Loss of Time Synchronization

The performance of the designed wavelet is compared and contrasted with several known wavelets by means of computer simulations. We have designed a ranging/communication system with DQPSK modulation and 128 orthogonal subcarriers, corresponding to wavelet packet tree of 7 stages. Guard intervals are not used and no error estimation or correction capabilities are implemented.

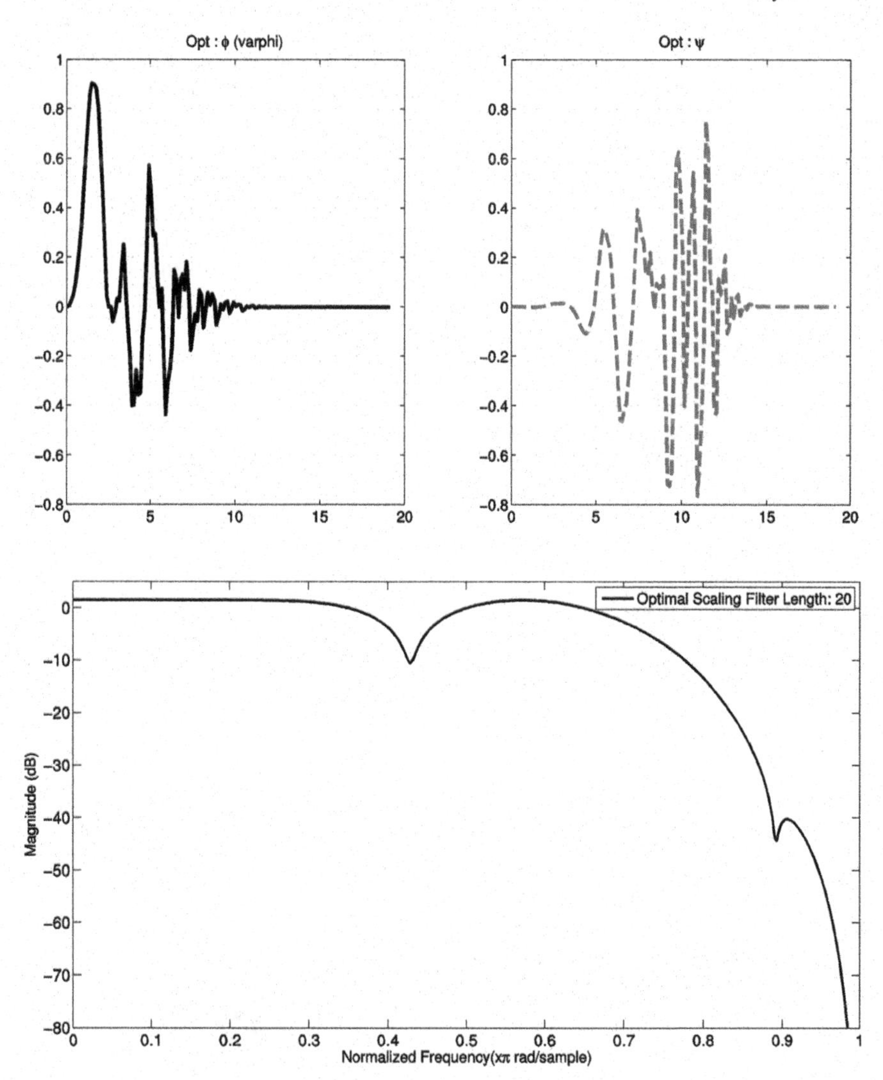

Figure 8.5 Optimal Filter; (a: top left) Scaling function, (b: top right) Wavelet function and (c: bottom) Frequency response in dB of the optimal filter.

To simplify the analysis, perfect frequency and phase synchronization are assumed. The time offset Δ_t is modeled as discrete uniform distribution between –2 and 2 samples, i.e., $\Delta_t \in [-2, -1, 0, 1, 2]$. In order to accentuate the difference in performances between various wavelets, an oversampling factor of 15 is applied. The details are tabulated in Table 8.3.

Table 8.2 Wavelet specifications and objective function

Name	Length	K-Regularity	$\sum_{n=0}^{\frac{L}{2}} (r_h[2n+1])^2$
Haar	2	1	–
Daubechies	20	10	0.41955
Symlets	20	10	0.41955
Discrete Meyer	102	1	0.45722
Coiflet	24	4	0.41343
Optimal	20	5	0.36814

Table 8.3 Simulation setup time synchronization error

	WPM
Number of Subcarriers	128
Number of Multicarrier Symbols per Frame	100
Modulation	DQPSK
Channel	AWGN
Oversampling Factor	15
Guard Band	–
Guard Interval	–
Frequency Offset	–
Phase Noise	–
Time Offset	Δ_t= 2

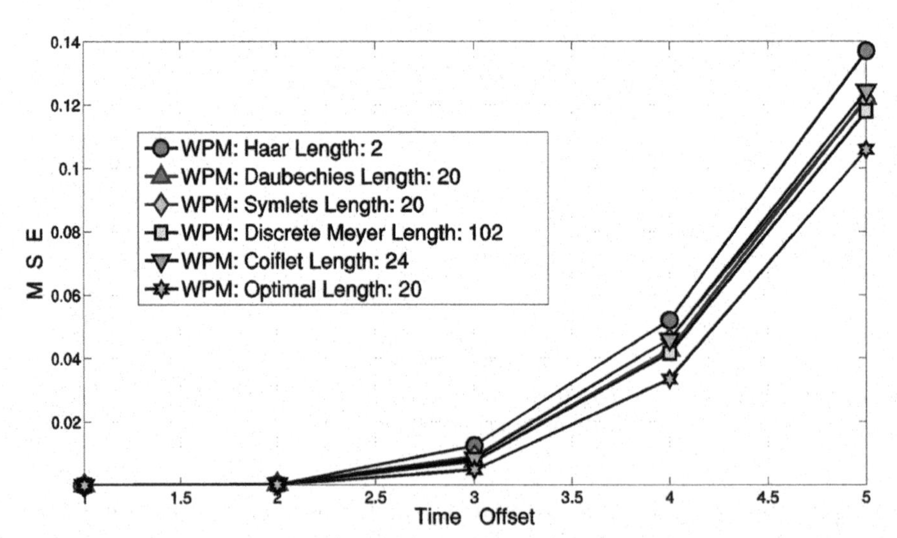

Figure 8.6 MSE of TOA estimation vs. Time Offset for WPM signal in AWGN channel (SNR = 20 dB).

The Mean Square Error (MSE) of TOA estimation is calculated for different values of time offset and is shown in Figure 8.6, respectively. Because the direction of timing error is inconsequential for WPM based system the time offset Δ_t is considered to follow a uniform distribution between 1 and 5 samples. The results presented corroborate the gains brought in by the newly designed wavelets.

8.7 Conclusions

Wavelet Packet Modulation is a strong candidate for signal design in 5G networks. It provides communication and ranging functionalities with one technology and offers enormous adaptability and flexibility to system designers. This tutorial chapter presented a methodology for designing new wavelets for communication and ranging of context-aware 5G networks when there is a time synchronization error. The design process was described as an optimization problem that accommodated the design objectives and additional constraints necessary to ensure the wavelet existence and the orthonormality. In order to obtain the global minimum, the original non-convex constraints and the objective function were translated into the autocorrelation domain. Using the new formulation, the design problem was expressed as a convex optimization problem and efficiently solved using the semi definite programming technique. To demonstrate the design mechanism for the ranging applications special filters were developed for the TOA estimation when there exists a time synchronization error. The simulation results revealed that the newly designed wavelet signal satisfied all the design objectives and outperformed the standard wavelets in terms of the MSE of TOA estimation. Importantly, the wavelet design framework presented in this chapter can easily be applied to other design criteria of 5G networks (e.g., security, spectral efficiency, throughput, latency performance, . . .) by merely altering the objective function. However, to be able to do so, the desirable properties of the wavelet bases must be translated into realizable objective functions. This can at times be challenging because the relationship between wavelet functions and filters is implicit and not direct.

References

[1] M. Benedetto, Understanding Ultra Wide Band: Radio Fundamentals, Prentice Hall PTR, NJ. 2004.

[2] M. Ghavami, L. B. Michael and R. Kohno, Ultra Wideband Signals and Systems in Communications Engineering, Wiley, 2004.

[3] Y. T. Chan, K. C. Ho, “A Simple and Efficient Estimator for Hyperbolic Location,” IEEE Transactions on Signal Processing,” Vol. 42, No. 8, pp. 1905–1915, Aug. 1994.
[4] H. Nikookar and R. Prasad, Introduction to Ultrawidebband for Wireless Communications, Springer, 2009.
[5] H. Nikookar, Wavelet Radio: Adaptive and Reconfigurable Wireless Systems based on Wavelets, Cambridge University Press, 2013.
[6] Daubechies, “Ten Lectures on Wavelets”, Society for Industrial and Applied Mathematics, Philadelphia, Pennsylvania, 1992.
[7] P. P. Vaidyanathan, Multirate Systems and Filter Banks, Prentice Hall PTR, Englewood Cliffs, New Jersey, 1993.
[8] M. Jansen and P. Ooninc x, Second Generation Wavelets and Applications, Springer, 2005.
[9] W. J. Phillips, “Wavelet and filter banks course notes”, 2003 [Available Online] http://www.engmath.dal.ca/courses/engm6610/notes/notes.html
[10] M. Vetterli and I. Kovacevic, Wavelets and Subband Coding. Englewood Cliffs, New Jersey: Prentice-Hall PTR, 1995.
[11] E. L. Lawler and D. E. Wood, “Branch-and-bound methods: A survey”, JSTOR Operations Research, Vol. 14, No. 4, pp. 699–719, August 1966.
[12] E. Alizadeh, “Interior point methods in semidefinite programming with applications to combinatorial optimization”, SIAM Journal of Optimization, Vol. 5, pp. 13–51, 1995.
[13] Y. Ye, Interior Point Algorithms: Theory and Analysis, New York, Wiley, 1997.
[14] S. Boyd, L. El Ghaoui, E. Feron, V. Balkrishnan, “Linear matrix inequalities in system and control theory”, SIAM Study in Applied Mathematics, Vol. 15, June 1994.
[15] S. Boyd, L. Vandenberghe, Convex Optimization, Cambridge University Press, 2004.
[16] L. Vandenberghe, S. Boyd, “Semidefinite programming”, SIAM review, Vol. 38, No. 1, pp. 49–95. March 1998.
[17] H. Wolkowicz, R. Saigal, L. Vandenberghe, Handbook of Semidefinite Programming, Kluwer Academic Publisher, 2000.
[18] R. Hettich, K. O. Kortanek, “Semidefinite programming: Theory, methods and applications”, SIAM Review, Vol. 35, No. 3, pp. 380–429, September 1993.
[19] A. Karmakar, A. Kumar, R. K. Patney, “Design of an optimal two-channel orthogonal filterbank using semidefinite programming”, IEEE Signal Processing Letters, Vol. 14, No. 10, pp. 692–694, October 2007.

[20] J. Wu, K. M. Wong, "Wavelet packet division multiplexing and wavelet packed design under timing error effects", IEEE Transaction. on Signal Processing, Vol. 45, No. 12, pp. 2877–2890, December 1997.
[21] J. K. Zhang, T. N. Davidson, K. M. Wong, "Efficient design of orthonormal wavelet bases for signal representation", IEEE Transactions on Signal Processing, Vol. 52, No. 7, July 2007.
[22] H. Nikookar, "Signal design for context aware distributed radar sensing networks based on wavelets," IEEE Journal of Selected Topics in Signal Processing, Vol. 9, No. 2, March 2015.
[23] B. Alkire, L. Vandenberghe, "Convex optimization problems involving finite autocorrelation sequences", Mathematical Programming, Series A93, pp. 331–359, 2002.
[24] T. N. Davidson, L. Zhi-Quan, K. M. Wong, "Orthogonal pulse shape design via semidefinite programming", IEEE International Conference on Acoustics, Speech and Signal Processing (ICASSP '99), Vol. 5, pp. 2651–2654, March 1999.
[25] T. N. Davidson, L. Zhi-Quan, J. F. Sturm, "Linear matrix inequality formulation of spectral mask constraints with applications to FIR filter design", IEEE Transaction. on Communications, Vol. 50, No. 11, pp. 2702–2715, November 1999.
[26] J. F. Sturm, "Using SeDuMi 1.02, a MATLAB toolbox for optimization over symmetric cones", Optimization Methods and Software, Vol. 11–12, pp. 625–653, 1999. [Available Online] http://sedumi.mcmaster.ca/
[27] J. Löfberg, "A Toolbox for modeling and optimization in MATLAB", Proceedings of the CACSD Conference, 2004. [Available Online] http://control.ee.ethz.ch/~joloef/wiki/pmwiki.php

Appendix: Sum of squares of cross-correlation

The sum of squares of cross-correlation magnitude is related to the autocorrelation sequences of low pass filter H and high pass filter G according to the following equation:

$$\sum_{n=0}^{L-1} |r_{hg}[n]|^2 = \sum_{n=0}^{L-1} r_h[n]\left((-1)^n r_h[n]\right)$$

$$= r_h[n] \cdot r_g[n] \tag{8.A1}$$

Proof

$$\begin{aligned}
\sum_{n} |r_{hg}(n)|^2 &= \sum_{n} \left(\sum_{p} h[p+n]\, g[p] \right)^2 \\
&= \sum_{n} \sum_{m} \sum_{p} h[p+n]\, g[p]\, h[m+n]\, g[m] \\
&= \sum_{m} \sum_{p} g[p]\, g[m] \sum_{n}^{L} h[p+n]\, h[m+n] \\
&= \sum_{m} \sum_{p} g[p]\, g[m] \sum_{n=m-p} h[m]\, h[2m-p] \\
&= \sum_{m} \sum_{p} r_h[m-p]\, g[p]\, g[m] \\
&= \sum_{p} \sum_{n=m-p} r_h[n]\, g[p]\, g[n+p] \\
&= \sum_{n} r_h[n]\, r_g[n] \\
&= r_h[n] \cdot r_g[n] \qquad (8.A2)
\end{aligned}$$

About the Author

Homayoun Nikookar received his Ph.D. in Electrical Engineering from Delft University of Technology in 1995. In the past he has led the Radio Advanced Technologies and Systems (RATS) program, and supervised a team of researchers carrying out cutting-edge research in the field of advanced radio transmission. He has received several paper awards at international conferences and symposiums. Dr. Nikookar has published about 150 papers in the peer reviewed international technical journals and conferences, 12 book chapters and is author of two books: Introduction to Ultra Wideband for Wireless Communications, Springer, 2009 and Wavelet Radio, Cambridge University Press, 2013.

9

TV Broadcast and 5G

Lars Kierkegaard

Teracom A/S, Copenhagen, Denamrk

9.1 Introduction

The traditional TV broadcast value chain is today subject to significant pressure due to new OTT market players such as Netflix and HBO entering the market. Traditional players are being by-passed by new market players in the value chain. Advanced 4G- and 5G mobile networks is expected to accelerate the pressure on the existing value chain even further, with LTE-Broadcast (eMBMS) being the disruptive trigger point. This chapter provides an overview of TV broadcast and 5G with outset in a historical perspective. In particular, the changes to the traditional TV broadcast value chain are described together with key technology drivers, and a view of TV broadcast in a future 5G scenario is presented.

This chapter focuses on TV broadcast and 5G. My reason for choosing this topic is not only the fascinating aspects of the technological development of Information and Communication Technology (ICT) but also the mere fact that television takes up a lot of our time. Just as an example, the average TV viewing in Denmark is three hours every day. Hence, it is clear that television plays a significant part of our lives.

The first part of this chapter looks at TV broadcast in a historical context. In order to reflect on TV Broadcast in a 5G era it is important to understand television in a historical context and the traditional TV broadcast industry, as we have known it for many years. I will walk you through the main milestones of the technological development in Denmark as well as the traditional TV broadcast value chain.

The second part focuses on the disruptive changes which the TV broadcast industry is undergoing these years. In fact, television as we have known it since the 1940's is subject to enormous changes caused by ground breaking developments in ICT technology and subsequent changes in viewing habits especially from the younger part of the population.

The internet, and especially IP and WiFi have resulted in an enormous disruptive change to the traditional broadcast TV value chain. New players have entered the value chain and existing players, especially the distributors of TV signals, are under pressure from Internet Service Providers who distributes the TV signals at lower costs mainly due to economies of scale and standardization.

9.2 Traditional TV Broadcast

The Figure 9.1 shows an American family watching TV in the living room in 1958. The picture illustrates the fact that there was only one television in the home and that it was something that the family gathered around to watch together. In a Danish context, public TV broadcast took its beginning in 1951 when Statsradiofonien started to transmit regular TV broadcasts each lasting one hour, three days per week, to only approximately 200 TV receivers.

In the following years, eight TV transmitter towers were established giving nationwide coverage for public service TV broadcast in Denmark in an analogue terrestrial TV broadcast network. In 1987, a competition public

Figure 9.1 An American family is watching TV in 1958. The family was gathered around the same screen [3].

broadcaster, namely TV 2, went live on air, following a political decision that Danmark's Radio should no longer have a monopoly on broadcast TV. This expanded the transmitter infrastructure to 34 main towers, which we also have in Denmark today.

In the 1980's the Danish government decided to liberalize, a part of the Danish TV broadcast industry and open up for commercial TV broadcasters when cable-TV and satellite-TV access technologies became available. The next major milestone in Danish TV history took place on 2 November 2009 when the analogue terrestrial TV network was shut down and the Digital Terrestrial TV (DTT) network based on DVB-T1 was launched 15 minutes later. A part of the network was later upgraded to DVB-T2. All that was needed was a digital set top box, which connects to the TV. At the same time, Boxer TV went live on air as a new DTT commercial pay TV provider in Denmark. Suddenly, it was possible to view more than 45 TV-channels on the terrestrial TV network.

Having introduced the main historical milestones let us look at what we could call the traditional TV broadcast value chain. The value chain is illustrated in Figure 9.2.

In the first part of the value chain, we have the TV broadcasters, who is responsible for production of their own content and who buys content from other content owners. The TV broadcasters can be either public service TV broadcasters such as DR, or pay-tv providers such as Boxer-TV or WAOO! In Denmark. The content is arranged and divided into TV channels, which is transmitted to the viewers via different distribution platforms.

The distribution network technologies include the DTT platform, satellite, Cable TV, DSL using copper cables and fibre. The third part of the value chain is the receiving end equipment, namely the televisions and set top boxes supporting either DVB-T/T2 for the DTT network or DVB-C for fiber based networks. The last part of the value chain is the consumers who receive the

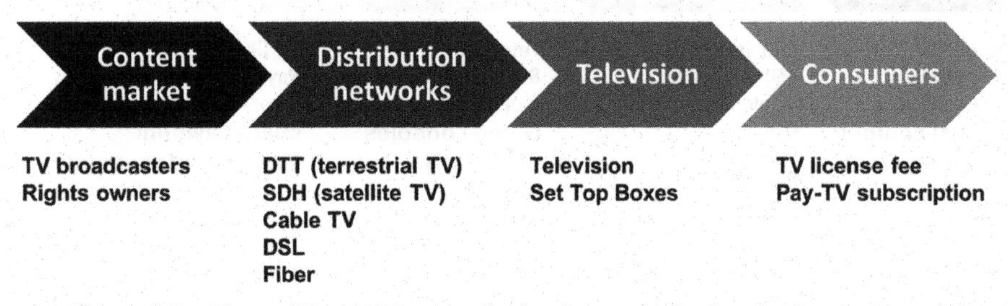

Figure 9.2 The traditional TV value chain as it have looked until a few years ago.

content and pays a TV license fee for the Public Service content and as an option pay-tv subscriptions for the pay-tv content.

9.3 Disruptive Changes

The Figure 9.3 illustrates the new TV value chain. Notice that new players have entered the different parts of the value chain. The traditional TV broadcasters now also face competition from on-demand content aggregators and Over The Top (OTT) providers such as Netflix and HBO. The traditional television is today only one of many devices from which it is possible to watch TV. This includes smart phones, tablets and game consoles, which now support TV viewing via apps. The traditional TV itself has been "internet enabled", which is also called smart TV's, and supports a long range of apps for the user.

Thus, new players have entered the market, and the market has become more fragmented with new roles. Examples of new roles in the value chain are OTT-providers, application developers, new device types and the fact that the consumer can create his or her own "video" channel (e.g., on YouTube).

The main reason for this disruptive development in the TV market is the convergence or melting together of different industries. Convergence in the TV market is a term used to describe the melting together of internet, broadcast and telecommunications industries. IP and internet is an enabler, which breaks down industry barriers, as it has been the case for other technological innovations during recent years. Refer to Clayton Kristensen's famous book on innovation for other examples on disruptive innovations. New business models and market players emerge and incumbent players in the value chain are challenged. The market also becomes more fragmented.

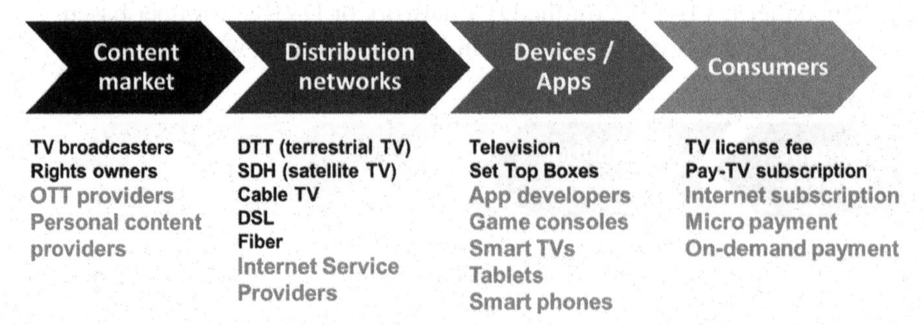

Figure 9.3 The disrupted TV broadcast value chain with many new players entering the market. New player categories are indicated with red text.

Historically, internet, broadcast and telecommunications have been separate and isolated industries. However, all of this is changing now with new players entering the market. Netflix who entered the Danish TV market in the autumn of 2012 with their Internet TV offering is an example of this. Netflix have entered the TV market with big strength not only in the Danish TV market but also on a global scale. Netflix provides a stand-alone OTT service, which operates across any Internet Service Providers network, whereas HBO offers their OTT TV offering as a valued added service to Telia Broadband.

Via Play, TDC Play and TV2 Play are examples of other new OTT streaming apps where broadcasters stream TV over IP directly to the consumer and are thus bypassing traditional distribution links in the TV value chain, such as terrestrial distribution, cable and satellite.

However, new hybrid standards such as Hbb-TV have also emerged. The Hybrid solutions combine broadcast and point-to-point IP communication, where broadcast transmits TV signals downlink to the consumers and internet is used uplink to provide new interactive services such as personalized services based on metadata. Examples of metadata are personalized electronic program guides where program highlights are presented based on the consumer's interests and preferences as well as predefined language and audio settings again according to the consumer's preferences.

9.4 Technology Drivers

IP and the internet are together with new wireless access technologies the main technology drivers behind the disruptive change in the TV broadcast value chain. Flow TV is just only one of many services and applications, which will be supported from the same network infrastructure in the future. In order words: ICT is becoming a utility just like electricity, water and energy. The main technology drivers are described in the following.

9.4.1 IP-based Communication

Common IP-based ICT infrastructure networks bring significant advantages for the operators and the users compared to digital proprietary distribution platforms. Firstly, all components in an IP-based network can be supervised and remote controlled from central locations. This is possible because all components have a unique IP address in the system, which also is the prerequisite for Internet of Things. In other words, supervision, operations and maintenance of the network is centralized, which drives down operating costs to a lower level compared with analogue or digital proprietary systems. For this reason, the Network Operation Center (NOC) plays a central role with a long range of remote services.

Figure 9.4 shows an example of a TMA based proprietary digital switch from 1999 used for coastal emergency radio voice communication between a control room and ships in distress. Notice the left picture with the cross-connects, which often require physical measurements to troubleshoot. The entire switch fills up more than a 40 square meter room.

The next picture in Figure 9.5 shows an IP-based "exchange" supporting the same coastal radio service as the proprietary digital exchange shown in Figure 9.4.

Figure 9.4 Digital exchange for ship-to-shore radio anno 1999 in Denmark [6].

Figure 9.5 IP-based digital exchange for ship-to-shore radio anno 2016.

The difference in size, power consumption and ultimately cost speaks for itself. Troubleshooting is done using software based protocol tracing tools in different layers of the OSI stack. The cost is also substantially lower because Commercial Off-The-Shelf (COTS) hardware is used.

9.4.2 Broadcast vs. Point-to-Point Communication

Today, the main difference between broadcast networks and present 3G (UMTS) and 4G (LTE Advanced) cellular mobile network, is the fact that broadcast networks transmits the same information to all users, whereas cellular mobile networks use point-to-point communication. The main advantage with broadcast networks compared with 3G and 4G mobile networks is the fact that the capacity requirements in the networks are the same no matter how many users receive the transmitted signal. In other worlds, it does not matter if two thousands or two million viewers are receiving the signal.

The disadvantage is the lack of interactivity, which is on the other hand, is the key advantage in point-to-point communication. In point-to-point communication networks the issue is that the capacity requirements are dependent on the number of concurrent users connected to the network.

9.4.3 LTE-Broadcast

The disadvantage of point-to-point communication compared to broadcast with respect to capacity is the main reason for the introduction of LTE-Broadcast, or also called eMBMS, in the 3GPP specifications. LTE-Broadcast was initially introduced in release 9 of the 3GPP specifications and further refined in releases 10 and 11.

However, to date only KT in Korea has commercially launched an LTE-Broadcast service. There is approx. ten trials in different parts of the world according to Global mobile Suppliers Organization (GSA).

9.4.4 Data-rates and Bandwidth

Bandwidth efficiency has increased significantly in 4G compared to 3G since IP communication is also used in the wireless access part of the mobile network. This improves the frequency band more efficiently compared to traditional TDMA, FDMA and CDMA coding schemes used in 2G and 3G networks. Thus, new and enhanced coding and modulation schemes have improved the bandwidth efficiency.

Table 9.1 Technologies and bandwidths for 2G, 3G, 4G and 5G technologies

Generation	Year	Standards	Technology	Bandwidth	Data-rates
2G	1991	GSM, GPRS, EDGE	Digital	Narrowband	<80–100 Kbit/s
3G	2001	UMTS, HSPA	Digital	Broadband	<2 Mbit/s
4G	2010	LTE, LTE Advanced	Digital	Mobile broadband	xDSL like experience 1 hr HD mobile in 6 minutes
5G	2025–2030	–	Digital	Ubiquitous connectivity	Fibre like experience 1 hr HD mobile in 6 seconds

The Table 9.1 shows the technologies and bandwidths for 2G, 3G, 4G and 5G networks respectively.

Notice the exponential increase in data-rates in the developing generations of mobile networks. The increase in data-rates in 5G networks in mainly due to the use of higher frequencies above 3 GHz with limited geographical coverage of 50–500 m in each cell. This concept is also called Small Cells.

Flow-TV and On-demand video requires that high bandwidths are supported in the access network and for backhaul connectivity. A main prerequisite of the increased data-rates in 5G networks is the availability of spectrum allocated to 5G networks.

At the ITU World Radio Conference (WRC) in Geneva in 2012, it was decided to reallocate the 700 MHz band ranging from 694–790 MHz from DTT to mobile internet. The main argument for this reallocation was the availability of enhanced codecs video compression in DTT networks such as HEVC. The other argument was that the 700 MHz band is more suitable for rural areas in Denmark, due to the longer geographical reach of this frequency band compared to other frequency bands, which are currently available for mobile internet. At WRC 2015, it was decided to postpone a proposal for the sub-700 MHz band to 2023.

The sub-700 MHz band is currently reserved for downlink TV-broadcast. Hence, it can be expected that the current DTT distribution platform will remain in service at least until 2030 in most countries including Denmark.

9.5 TV Broadcast in the 5G Era

One of the main characteristics with 5G networks is the use of higher frequencies over 3 GHz which have limited coverage from approx. 50–500 meters from the base station antenna. 5G networks will support bandwidths of up to 10 Gbit/s. TV everywhere is a term used to describe:

- Freedom of choice to only pay for TV-channels that the consumer wants to watch.
- Freedom to watch the TV-channels irrespective of the time.
- Freedom to watch the TV-channels irrespective of place (at home, commuting, in the summerhouse etc.

The in 5G area these capabilities are mandatory for the consumer. A significant part of these capabilities will most likely be supported in future 5G networks.

However, security is a major concern one major when it comes to TV broadcast using LTE-Broadcast in 5G networks is security which needs to be addressed is security. Adequate security measures are critical in terms of securing service continuity, service integrity and service confidentiality. TV viewers demand a stable and robust TV-broadcast service and they have an increasing demand to have the privacy in terms of preferences and choices of content. Cybercrime is becoming an increasing issue on IP- and internet connected systems.

Thus, the future ICT communication landscape for TV broadcast should not only consist of both IP and internet based content but also non-IP digital TV broadcast platforms that cannot be hacked and which continues as a robust and reliable communication means to the population both in everyday life and in case of major crises. As an example, a major fire occurred outside of Stockholm in 2014. The internet in the area was down for three days but the TV and radio broadcast networks continued to broadcast important information and messages to the population during the crises.

To conclude, a combination of both IP- and non-IP TV-distribution platforms for flow-TV are to be expected even in the 5G era, which not only serves to entertain but also to be an important communication means for the society in case of major crises in the society. Well, at least until 2030.

9.6 Conclusions

This chapter has analyzed the disruptive changes to the traditional TV-value chain, which has occurred in recent years. No doubt, IP and the internet have affected many industries, and TV broadcast is no exception. In addition, new market players have entered into the value chain, which increases competition but also new revenue opportunities being the end-result.

However, although the internet has changed the TV broadcast industry for good it is important to note that the IP and the internet and traditional

DTT-broadcast should be regarded as complementary platforms in order to secure a reliable and robust TV broadcast service.

Thus, a combination of both IP- and non-IP TV-distribution platforms for flow-TV are to be expected even in the 5G era, which not only serves to entertain but also to be an important communication means for the society in case of major crises in the society. Well, at least until 2030, as the former Director-General of the World Trade Organization, Pascal Lamy, also recommends in a report to the EU commission [11].

What happens after 2030 is difficult to say at the point of writing this chapter. No doubt, it will take time to get the security aspects solved with respect to TV-broadcast over IP-based networks and the internet with the increases in cybercrime. However, there is no doubt that in the longer run even more frequencies, especially in the higher frequency bands above 6 GHz, are likely to be allocated to 5G networks. This will in turn increase the capacity of OTT viewing even further, and especially when combined with the concept of Content Delivery Networks (CDN) meaning that the OTT content is located close to the end user, which in turn reduces load on the mobile network.

So to conclude: Only time will tell how TV-broadcast is distributed after 2030. However, one thing is for sure. Adequate ICT infrastructure is an important prerequisite in order to move viewing of moving images to the next level.

References

[1] DR Report, DR Medieforskningen 2015.
[2] da.wikipedia.org/wiki/Fjernsyn_i_Danmark
[3] en.wikipedia.org/wiki/Television_in_the_United_States
[4] www.teracom.dk
[5] Clayton Kristensen: The Innovators Dilemma, Havard Business School Press, 1997.
[6] Photograph source: Lars Kierkegaard.
[7] Photograph source: Lars Kierkegaard.
[8] www.3gpp.org
[9] GSA: Evolution to LTE Report, October 2015.
[10] EU Commission Report, Mobile Communications: From 1G to 5G.
[11] Pascal Lamy: Report to the European Commission, Results of the work of the High Level Group on the Future use of the UHF band (470–790 MHz), 2014.

About the Author

Lars Kierkegaard is currently Head of Strategy & Business Development at Teracom A/S. Lars is an expert on ICT and network convergence and have a 17 years background from international hi-tech ICT companies within media, broadcast, telecom and public safety industries.

He joined Teracom in 2010 where he was responsible for standardization of existing technical broadcast solutions into market oriented ICT product and service portfolios.

In 2013, Lars was appointed with the responsibility to head a new business innovation department to drive growth initiatives. At the same year, Lars was awarded with the title "Employee of the year" in TeracomBoxer Group in the category "Profitable Growth in our whole business".

In 2015 Lars, lead Teracom to win a 15-year contract for design, system integration and operations of a new nationwide coastal radio system in Denmark.

Main responsibility areas include business development & sales, people management, product management, innovation, and bid management. He is also a frequent speaker on ICT and convergence at international conferences.

Prior to his employment at Teracom, he has held senior product management and business development positions in Terma, Anritsu and Ericsson with Information and Communication Technology (ICT) being the "red thread".

Lars is board member of the Danish Consumer Electronics Association. He is also advisory board member at Center for Communication, Media & ICT at Aalborg University, board member of the IEEE Joint Chapter for Denmark, and external censor at the technical universities in Denmark. Finally, Lars is board member of the WorldDAB Association.

He holds a M.Sc. degree in Civil Engineering from the Technical University of Denmark, and a Master's degree in Information and Communication Technology from Aalborg University in Denmark, as well as a Graduate Certificate in Business Administration from Copenhagen Business School.

10

The Next Mobile Communication Steps into New Application Areas

Walter Konhäuser

Technical University of Berlin, Germany

10.1 Introduction and Mobile Pioneer Phase

The digital mobile communication started 1992 with first GSM network deployments in Europe, called 2nd Generation. Meanwhile we are discussing 5th Generation networks planned to start first deployments in 2020. This chapter describes the mobile evolution and its highlights. It will consider 5G technologies focus on new application areas which are not yet covered by mobile communication systems. As an example for new application areas it will describe a decentralized energy storage system and its mobile communication requirements to make renewable energy supply viable. The chapter ends with Conclusions and outlook.

Mobile communication is older than many today's users would expect. During the first period, called the pioneer phase from 1921 to 1945, mobile radio technology was still quite premature and dominated by the mechanical problem designing a radio system that could survive the bumps and bounces of a moving vehicle. These initial radio systems were based on amplitude modulation radio and no commercial services were established. The invention of frequency modulation was an important determinant and the World War II brought tremendous improvements in designs. This first pioneer phase is marked by three topics [1]:

- mainly used by military and paramilitary groups,
- building a radio transmitter capable of operating within the size and power constrains of a moving car and
- radio transmission based on frequency modulation.

The next phase of development of mobile communication systems is the expansion of mobile telephone services into commercial use. The technical improvements were oriented towards two goals:

- reduction of transmission bandwidths and
- implementation of automatic trunking.

The next step of the early mobile communication systems was

- cellular structuring.

This innovation represented a totally new approach of deploying radio networks. It was the idea to extend the capacity to an unlimited system, breaking through the limits that had restricted the growth of the mobile communication.

10.2 Analog Mobile Communication Systems

After World War II a lot of mobile communication systems based on analog technology were deployed in different countries. This was the start of the commercial phase, but still a niche market. These networks had following attributes:

- analog air channel,
- narrow band communication,
- voice centric services,
- only national roaming,
- FDD based,
- less subscriptions compared with today,
- expensive tariffs,
- used in cars mainly and

were called 1st Generation of mobile communication systems. Most common 1st Generation systems are AMPS, TACS, NMT and C-net.

10.3 Digital Mobile Communication Systems

A totally change in the mobile communication began with the development of first digital systems in the eighties of the last century with first deployments 1992. It started from a niche market with less subscriptions and high tariffs to a mass market with very high penetration and flat rates and remarkable changes of our daily human behaviour. It is in the meanwhile unimaginable to live without a mobile terminal. The most successful system of this 2nd Generation of mobile networks was GSM, which was planned as a mobile communication system for Europe only and conquered nearly the whole world in less than 10 years. In 2002, when the 3rd Generation started, 62% of the worldwide subscriptions were based on GSM (Figure 10.1). The reasons for this success were:

- fully digital transmission,
- enhanced frequency economy,
- privacy,
- improved hand-held viability,
- flexible configuration of networks and services,
- international compatibility,
- enhanced services.
- based on field-proven digital switch technology,
- open Interfaces,
- multi-vendor architecture,
- competitive operator concept and
- early system availability.

But nevertheless the killer application still was mobile voice service. 2nd Generation standards were narrow band systems with voice centric services and the first real data services started with the introduction of GPRS, what led

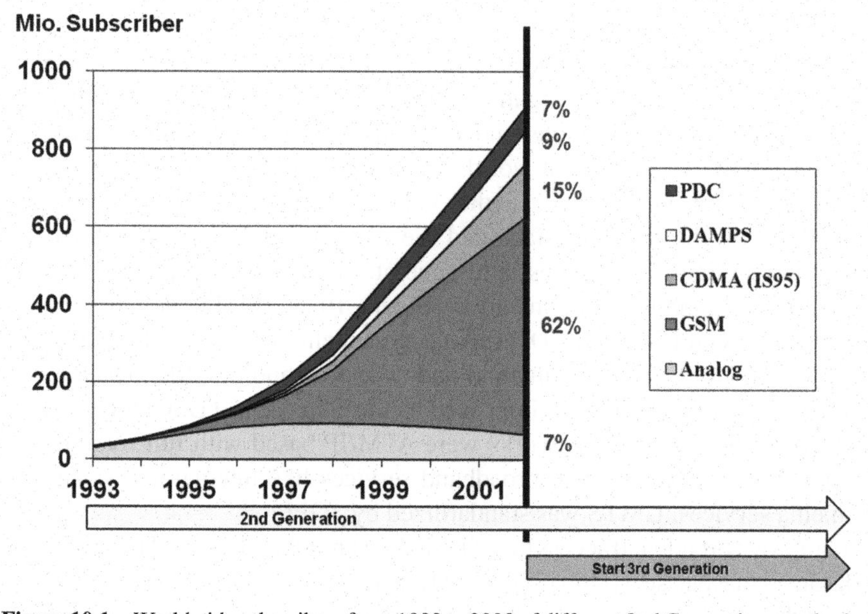

Figure 10.1 Worldwide subscribers from 1993 to 2002 of different 2nd Generation standards.

to an extended architecture. Another milestone was the introduction of the IN-architecture with prepaid service. The next step into broadband was the development of EDGE architecture with the success to become member of the IMT 2000 family. EDGE gives GSM/GPRS-Market new significant boost and business potential, EDGE and 3G WCDMA are complementary not competing and EDGE is the smooth migration path to 3G-services in existing spectrum. Although EDGE requires no hardware or software changes to be made in GSM core network, base stations must be modified. EDGE compatible transceiver units must be installed and the base station subsystem needs to be upgraded to support EDGE. New mobile terminal hardware and software is also required to decode/encode the new modulation and coding schemes and carry the higher user data rates to implement new services.

The International Telecommunications Union (ITU) launched a new framework of standards under the generic name of IMT-2000 to present the culmination of ten years study and design work to identify the 3rd Generation standards. The following key factors were established [2, 3]:

- high speed access, supporting broadband services such as fast Internet access and multimedia applications,
- flexibility and support of new kinds of services and
- offer of an effective evolutionary path for 2nd Generation existing wireless networks.

The market expectations were focused on more advanced services than voice and low data services considering merging three worlds which have operated so far independently the computer and data activities, the telecommunication and the audio and video content world. Multimedia applications use several services in parallel such as voice, audio, video data, e-mail, etc. and have to be supported by the radio interface and the core network. The main focus on this merged field of activities is to generate new business opportunities and multimedia mobile communication (Figure 10.2). UMTS was the first radio network with FDD and TDD radio access technologies. For FDD was chosen WCDMA as multiplex method and access technology for TDD was TD-CDMA [2]. TD-CDMA was improved by the Chinese to TDSCDMA and deployed in China. UMTS networks were ATM/IP based with full roaming through different networks and broadband services like internet access and multimedia services. UMTS was standardized by 3GPP.

To improve broadband access to get DSL-like broadband speeds HSPA was standardized in Release 5 (Downlink) and Release 6 (Uplink) of 3GPP. Mobile broadband moved a step faster with HSDPA delivering up to 14 Mbps in the

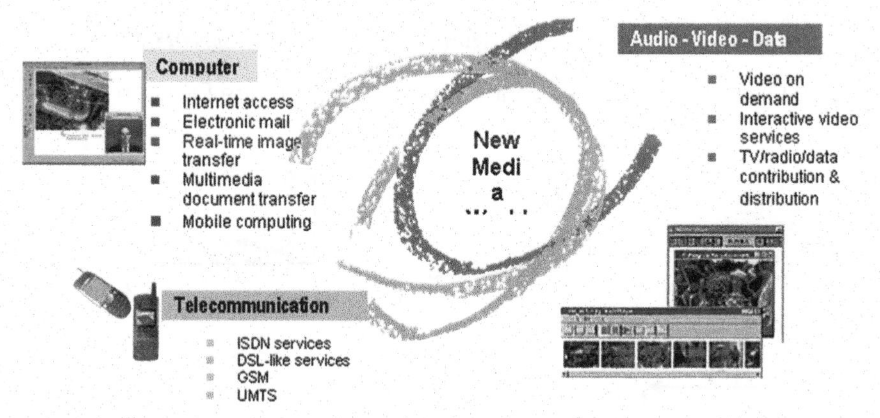

Figure 10.2 Converging three worlds generates new business opportunities (Picture Siemens).

downlink and HSUPA returning 5.8 Mbps in the uplink. HSPA improvements in UMTS spectrum efficiency were achieved through new modulation (16 QAM) formats, reduced radio frame lengths and new functionalities within radio networks (including retransmissions between Node B and the Radio Network Controller). Consequently, throughput is increased and latency is reduced (down to 100 ms and 50 ms for HSDPA and HSUPA respectively). By the end of 2007, there were 166 commercial HSDPA networks in 75 countries in operation and further 38 networks committed to deployment. The first commercial launch of HSUPA was in early 2007 and 24 networks had launched by the end of the year.

CDMA 2000, mainly developed in US and successor of the 2G standard IS 95, became member of the IMT-2000 family too. It is a hybrid 2.5G/3G standard of mobile communication that use CDMA, a multiple access scheme for digital radio, to send voice, data, and signaling data between mobile phones and cell sites. CDMA 2000 is considered a 2.5G protocol in 1xRTT and a 3G protocol in EVDO.

CDMA is a mobile digital radio technology that transmits streams of bits and whose channels are divided using pseudo noise sequences. CDMA permits many radios to share the same frequency channel. Unlike TDMA, a different technique used in GSM and DAMPS, all radios can be active all the time, because network capacity does not directly limit the number of active radios. Since larger numbers of phones can be served by smaller numbers of cell sites, CDMA-based standards have a significant economic advantage over TDMA-based standards, or the oldest cellular standards that used frequency division multiple access (Figure 10.3).

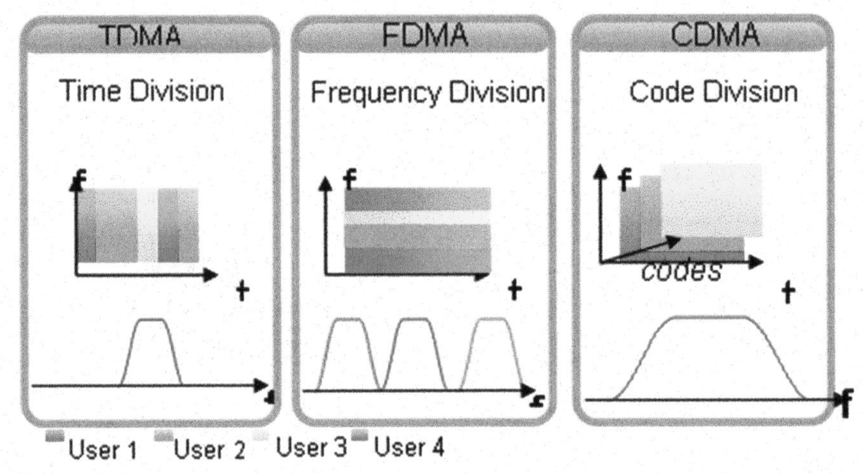

Figure 10.3 Major multiple access methods used in 2nd and 3rd mobile radio standards (Picture NSN).

CDMA 2000 remains compatible with the 2nd Generation CDMA telephony methods IS 95. The CDMA 2000 standards CDMA2000 1xRTT, CDMA2000 EV-DO, and CDMA 2000 EV-DV are approved radio interfaces for the ITU's IMT-2000 standard. CDMA 2000 is an incompatible competitor of the other major 3G standard UMTS. It is defined to operate at 400 MHz, 800 MHz, 900 MHz, 1700 MHz, 1800 MHz, 1900 MHz, and 2100 MHz. CDMA2000 was standardized by 3GPP2. IMT 2000 allocated worldwide news spectrums for 3G for FDD and TDD applications, except North America (Figure 10.4).

After deployment of HSPA and spreading of smartphones the mobile data traffic in the 3G networks increased extremely. Attractive HSDPA flat-rate mobile-broadband services offered to drive strong increase in traffic volumes. As traffic still grows faster than revenue, networks had to become more efficient. This required low cost per bit technologies and was one of the main arguments to push 4G (Figure 10.5).

LTE/SAE (Long-Term Evolution/System Architecture Evolution) is a standard for wireless cellular communication of high-speed data for mobile terminals. It is evolutionary based on the GSM/EDGE and UMTS/HSPA network technologies, increases the capacity and speed using a different radio interface together with core network improvements and decreasing the network costs /4,5/. The standard is defined by the 3GPP and is specified in its release 8 document series, with minor enhancements described in release 9.

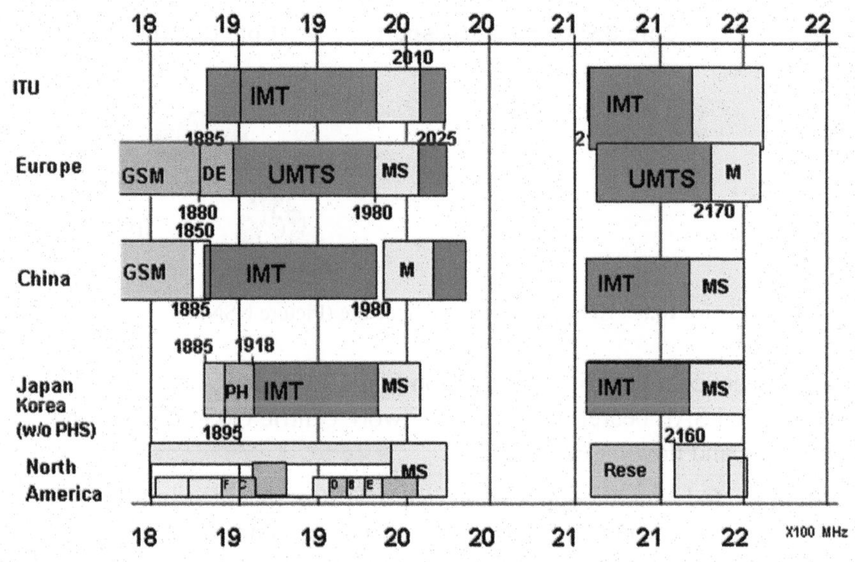

Figure 10.4 IMT 2000 worldwide frequency plans (Picture Siemens).

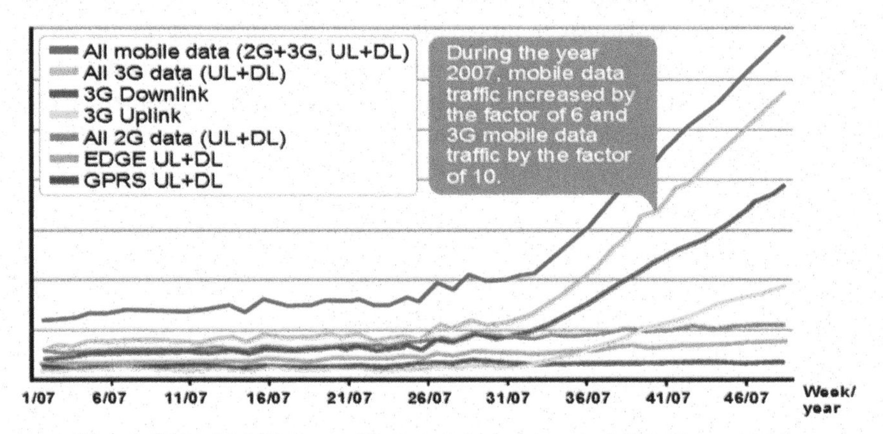

Figure 10.5 Mobile data traffic growth during the year 2007 (Picture NSN).

LTE is the natural upgrade path for carriers with both GSM/UMTS and CDMA 2000 networks. Only multi-band phones will be able to use LTE in all countries where it is supported. The key benefits for operators and users are:

- only one network element in radio and core each (Figure 10.6),
- enhancement of packet switched technology,
- high data rates, low latency, packet optimised flat IP system,
- comprehensive security,

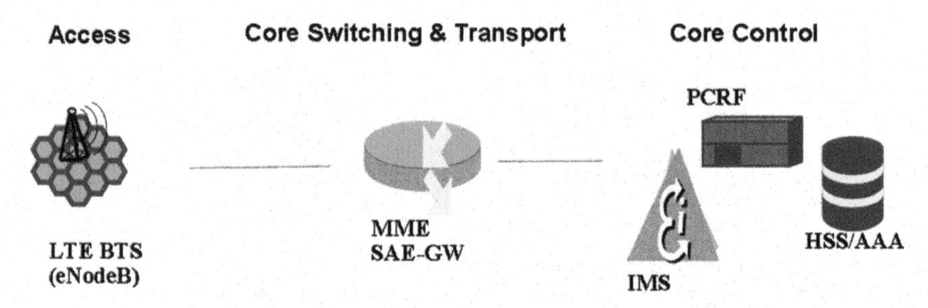

Figure 10.6 LTE/SAE system architecture (Picture NSN).

- mobility concept with tight Integration for 3GPP access,
- streamlined SAE bearer model with network centric QoS handling and
- on/offline and flow based charging.

Although marketed as a 4G wireless service, LTE (as specified in the 3GPP release 8 and 9 document series) does not satisfy the technical requirements the 3GPP consortium has adopted for its new LTE Advanced standard. The requirements were originally set forth by the ITU-R organization in its IMT Advanced specification. However, due to marketing pressures and significant advancements that WiMAX, Evolved High Speed Packet Access and LTE bring to the original 3G technologies, ITU later decided that LTE together with the aforementioned technologies can be called 4G technologies [4]. The LTE Advanced standard formally satisfies the ITU-R requirements to be considered IMT-Advanced [5–7].

- MME: Mobility Management Entity
- PCRF: Policy and Charging Control Function
- SAE-GW: System Architecture Evolution Gateway
- IMS: IP Multimedia Subsystem
- HSS: Home Subscriber Server
- AAA: Authentication, Authorization and Accounting

LTE radio interface use different methods in downlink and uplink. For downlink OFDMA with

- improved spectral efficiency,
- reduced interference,
- very well suited for MIMO and

for uplink Single-Carrier-FDMA with

- power efficient uplink increasing battery lifetime,
- improved cell edge performance by low peak to average ratio and
- reduced terminal complexity.

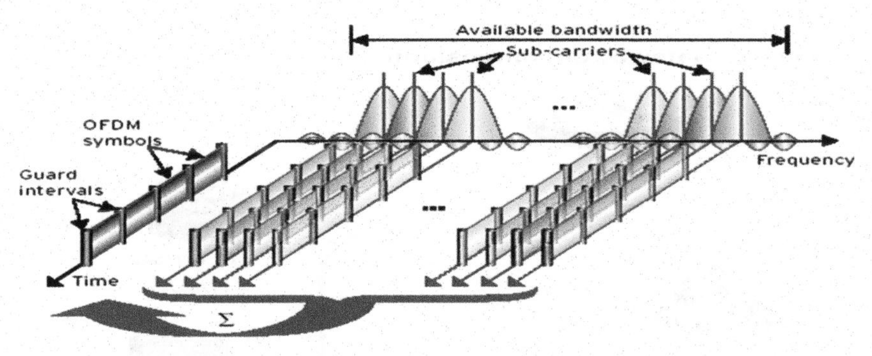

Figure 10.7 LTE radio principles (Picture NSN).

Further topics:

- enabling peak cell data rates of 173 Mbps downlink and 58 Mbps in uplink at 20 MHz bandwidth, FDD, 2 TX, 2 RX, DL MIMO, PHY layer gross bit rate
- scalable bandwidth: 1.4/3/5/10/15/20 MHz also allows deployment
- in lower frequency bands (rural coverage, refarming)
- short latency: 10–20 ms roundtrip ping delay (server near RAN)
- FDD and TDD timing.

Figure 10.7 illustrates the LTE radio principles. Core technology is based on the Mobility Management Entity with C-Plane part of a GW, session and mobility management, idle mode mobility management, paging and AAA proxy. The serving gateway contains user plane anchor for mobility between the 2G/3G access systems and the LTE access system and lawful interception. The packet data network gateway implies the gateway towards internet/intranets, user plane anchor for mobility between 3GPP and non-3GPP access systems like home agent, charging support, policy and charging enforcement, packet filtering and lawful interception (Figure 10.8).

The roadmap of the digital mobile network evolution towards 5th Generation, starting with first network operation based on GSM 1992 until today shows Figure 10.9.

10.4 5G Requirements and Technologies

Although 4G Technology is still in rollout a lot of research- and industry activities have started already. With 5G a new phase of mobile communication will start and the focus is on the billions of terminals available in the internet

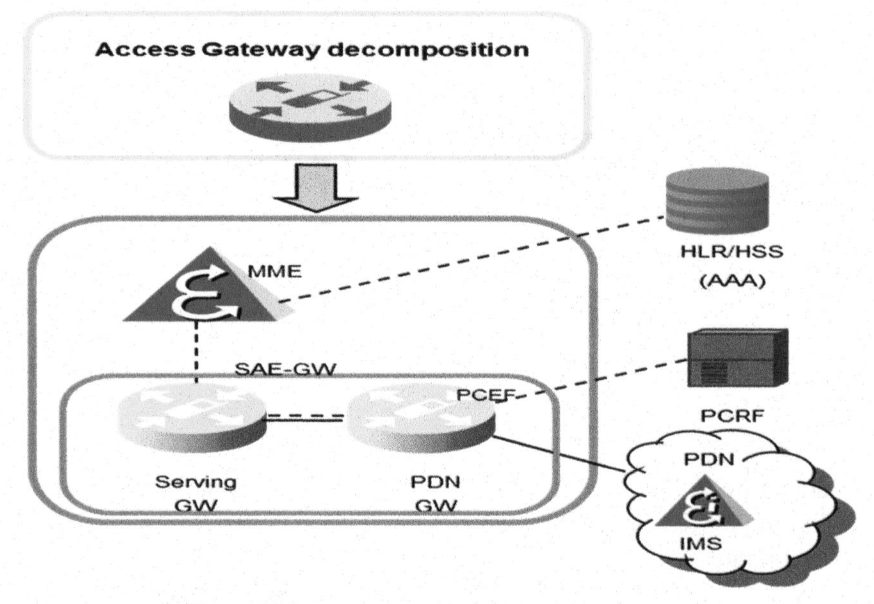

Figure 10.8 Core network principles (Picture NSN).

of things. We can, therefore, deduce that data traffic will continue to increase dramatically over time. It is estimated 1000 times by 2020. Connected devices will increase over time by 10–100 until 2020. New device types like probes, sensors, meters, machines, control computers etc. will significantly contribute to that increase. New sectors will bring new priorities, like future cars with car-to-car communication and real time navigation, Industry 4.0 with plant of the future, Energy with decentralized energy production and energy storage and real estate with metering, consumption control etc. Broadband communications will stimulatethe economy by contributing significantly to GDP and creating employment. From a user's point of view 5G networks should enable the perception of infinite capacity, tactile internet and augmented reality. Compared to existing networks, 5G networks will need to be more available, more dependable and more reliable and offer increased speed, increased throughput, decreased latency, improved deviceautonomy and must be offered at low cost. "5G is an end-to-end ecosystem to enable a fully mobile and connected society. It empowers value creation towards customers and partners, through existing and emerging use cases, delivered with consistent experience, and enabled by sustainable business models" is the NGMN 5G vision [10].

2. Generation
GSM

Voice, Data, GSM Ph. 2
Dual Band, Half Rate
basic services/
network optimisation
GSM 2+ and
Intelligent Networks(IN)
GPRS, EDGE

Goals:
Coverage/Capacity
Enhanced Services

3. Generation
UMTS

Wideband Air interface: WCDMA, TDSCDMA
Bandwidth on Demand, Seamless Services
FDD/TDD ATM/IP based Networks,
Enhanced Multimedia Services with full roaming through different networks,
HSDPA, HSUPA

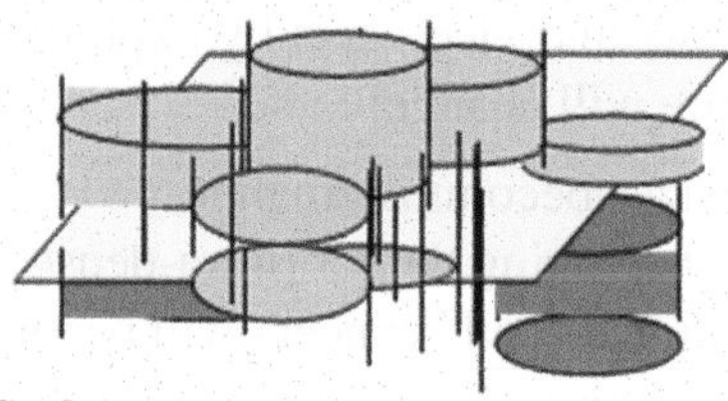

Goals:
New Business Opportunities
Multimedia Mobile Communication

4. Generation
LTE

Only one Network Element in
Radio and Core each
high data rates, low latency,
packet optimised flat IP system
Comprehensive Security
Broadband multipath radio
DL:OFDMA,
UL: Single-Carrier-FDMA
Datarates 1,4/3/5/10/20 Mbps

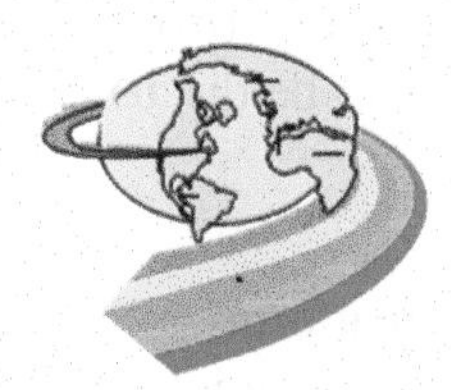

Goal:
Broadband Mobile IP-based Communication

5. Generation

Higher Data rates
Lower latency
Software Defined
Networking
Network Function Virtualization

Goal:
End-to-end ecosystem to enable a fully mobile and connected society

Figure 10.9 Digital mobile networks evolution towards 5th Generation.

The evolution process, which will be chosen, is of importance:

- embark on a linear evolution of today's networks as it was done in the past, or

- adopt a new revolutionary approach, or
- a combination of both.

Traditionally, the 3GPP approach is evolutionary (including backwards compatibility). A pure evolutionary approach may not be sufficient for 5G. To obtain new spectrum for mobile services is essential, but not the whole solution. More efficient use of spectrummust be ensured with licensed, unlicensed and shared access regimes. Opportunities to deliver broadcast content to mobile users and vice versa by exploiting synergies. Network topology and architecture will change:

- cell sizes becoming smaller,
- cells becoming denser (ultra-dense over time),
- traditional "cell" concept will become less relevant in favour of wireless cloud approach,
- interworking with other networks more prevalent (e.g., Wi-Fi offload),
- introduction of device to device working,
- virtualisation of network functions, management and orchestration,
- evolution from hardware to software.

From an economic point of view high demand to reduce CAPEX and OPEX. Decrease radio network energy consumption and CO_2 emissions. Reduce energy consumption per BTS site and use more energy efficient BTS, reduce site power consumption, increase site temperature, use outdoor BTS, optimize energy consumption versus traffic, reduce the number of BTS sites, increase cell coverage, share networks and use renewable energy source like solar cells, wind and hybrid solutions on site like green energy (hydro, wind and bio generated grid electricity). The technical issues to be discussed and analyzed:

- new chip technologies,
- new network architecture with distributed service platforms,
- new hierarchy of cloud platforms, e.g., mobile edge clouds for base stations,
- mini clouds on local level and less central clouds,
- new operating systems and network protocols,
- tactile internet,
- high availability,
- high reliability,
- high security,
- new diversity concept for frequency, place and infrastructure,

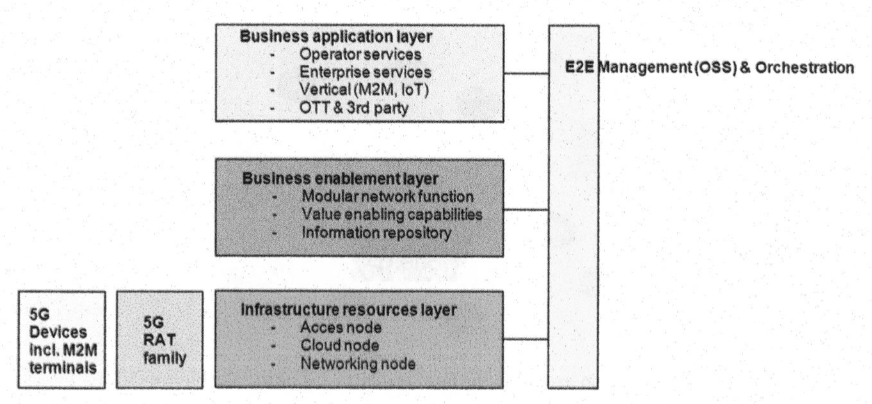

Figure 10.10 5G architecture envisioned by NGMN.

- future machine to machine communication with new authentication techniques to support applications with high security and real time requirements and
- new antenna technique with controlled and intelligent beaming.

Next Generation Mobile Network [8] created an architecture that leverages the structural separation of hardware and software, as well as the programmability offered by Software Defined Networking and Network Function Virtualization. 5G architecture is an SDN/NFV native architecture covering all aspects ranging from devices, (mobile/fixed) infrastructure, network functions, value enabling capabilities and all the management aspects to orchestrate the 5G system. The Infrastructure resource layer contains the physical resources of a fixed-mobile converged network, access nodes, cloud nodes, networking nodes and 5G devices with all kinds of terminals from vertical business applications like M2M terminals. The Business enablement layer contains all functions required within a converged network of modular network function, value enabling capabilities and information repository. The business application layer implies applications and services of the operator, enterprise services, verticals of M2M and IoT and utilisation of 5G networks by 3rd parties. E2E management and orchestration has the capability to manage this 5G network end-to-end and provide an operation support system and self-organizing network capabilities (Figure 10.10).

10.5 New Application Areas for Vertical Industries

The next phase of mobile communication is to integrate industry, energy, mobility, real estate, and its processes. This is widely referred to as machine

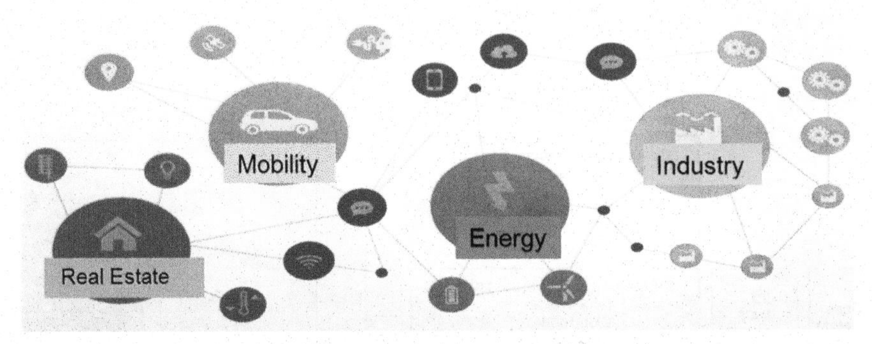

Figure 10.11 New application areas for vertical industries within 5G (Picture VDE).

to machine communication and Internet of the Things. Billions of smart devices will use their embedded communication abilities and sensors to act on their local environment and use remote triggers based on intelligent logic. These devices have different requirements with respect to capabilities, power consumption and cost. IoT will also have a wide range of requirements on networking such as reliability, security, performance (latency, throughput), etc. New services for vertical industries like

- future cars (car-to-car communication, real time navigation, etc.),
- industry 4.0 (plant of the future),
- energy (decentralized energy production, energy transport, energy distribution and
- energy storage) and
- real estate services (metering, consumption control) [8] (Figure 10.11)

will not be limited to connectivity only but can also require enablers from cloud computing, big data management, security, logistics, etc.

10.6 Application Example Decentralized Energy Storage

Energy is a very important resource for the benefit of mankind. In many countries discussions are ongoing how to manage the energy turnaround. An important contribution will build intelligent Smart Grid applications as new services within the communication networks. A decentralized energy storage concept based on Smart Grid applications and a centralized control system shows an example how to contribute to the energy turnaround. Merging power networks and communication networks to intelligent Smart Grids opens a lot of opportunities for future energy resource handling and should motivate researchers to find out the best solutions for the future [9]. For better monitoring and metering, sensors, communication technology as well as distributed real-time computing platforms will be the key technologies for managing various electrical equipment. The herewith handled state parameters are critical for the determination of the current state of the Smart Grids and seen as the success factor for shifting from centralized to decentralized energy system.

This captured information will be used as input for many types of model predictive algorithms whose output supports decisions to achieve the goals of the future Smart Grids. For an appropriate control structure in order to stabilize the grid with a limited mechanical inertia, an architecture based on an appropriate combination of central and decentralized control is needed.

For better predictive models and algorithms improved computer based models and algorithms will be needed. The grid elements such as transmission and distribution lines, voltage and current transformers, flexible AC and DC elements, switches, protection equipment but also all grid users including generators, storage, consumer equipment and behavior have to be extended with digital communication and information features.

A software architecture allowing consumers and market players to compose new services and to satisfy own requirements related to energy services and products thereby using also market interfaces and at the same time supporting the quality and security of supply of the grid based electricity system [10].

The energy turnaround initiated by politics created a high volatility in the energy production based on renewable energy (wind power stations and photovoltaic systems). To integrate renewable energy economically an additional concept of energy storage and energy supply has to be claimed. The proposed concept is a decentralized energy concept and decouples energy production from energy consumption. This is necessary to increase the share of renewable energy in the energy supply in an economical way (Figure 10.12). In the cellars of the real estate industry will be deployed thousands of battery systems. A battery system exists of a battery-farm, a AC/DC inverter and a control- and communication unit. All battery systems are linked to each other and to the control center via the communication unit called Smart Grid. Based on the Smart Grid all decentralized storage system can be linked to a virtual large storage system distributed countrywide. When there is an oversupply of energy the batteries can be loaded and when there is demand the batteries can be unloaded, controlled by the central control center. Intelligent data exchange

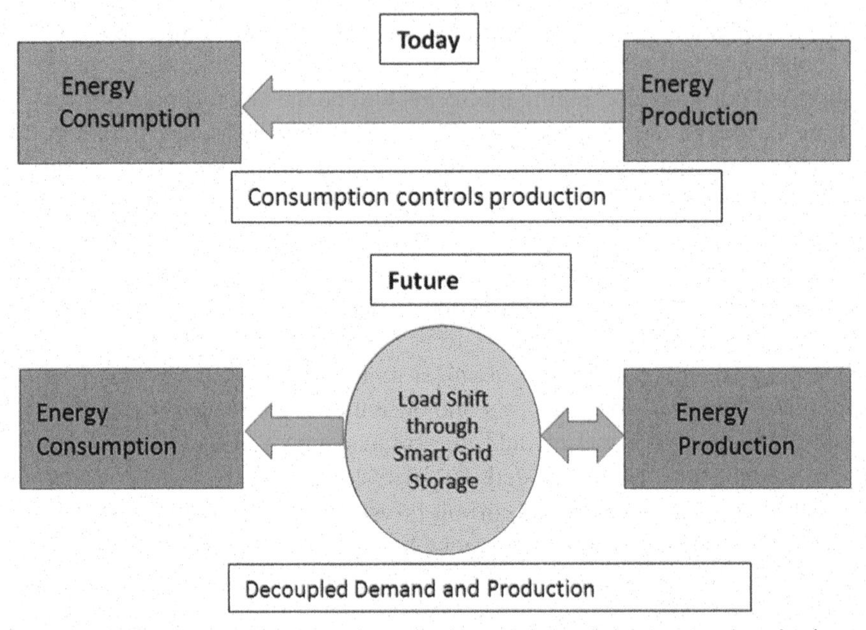

Figure 10.12 Consumption controls production and decoupled demand and production.

between the decentralized battery systems enables the battery systems to organize itself a lot of energy supplies tasks (Figures 10.13 and 10.14).

The Control- and Communication Platform (Figure 10.15) is composed of an industry PC, Linux operating system and the OGEMA Platform [11], developed by Fraunhofer IWES Institute. The applications have to organize

- control of the AC/DC inverter,
- control Battery Management System,
- communication with the central unit,
- communication with the battery systems belonging to the same Micro Grid,
- smart metering,
- recording balancing power,
- remote maintenance,
- security,
- IoT interface.

The revenue model is based on balancing power and tenant electricity. The focus is on real estate industry with huge leverage effect. The top 10 real estate

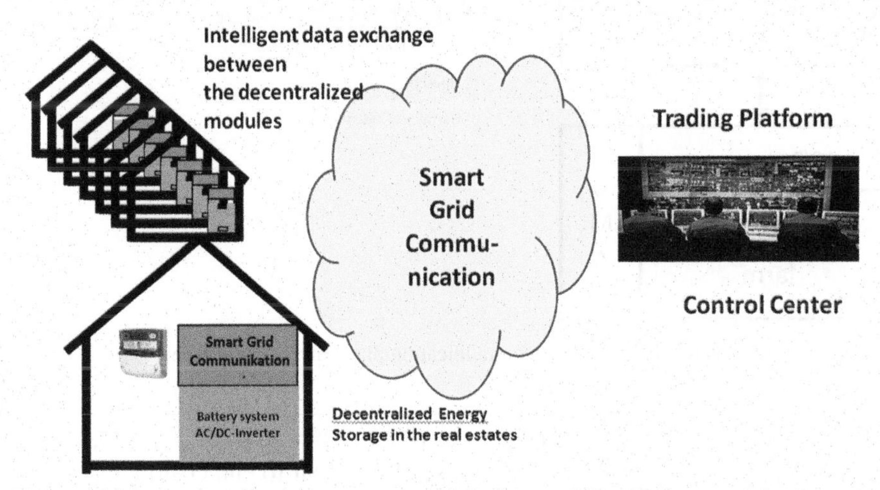

Figure 10.13 Decentralized energy storage linked on a Smart Grid communication to a control center.

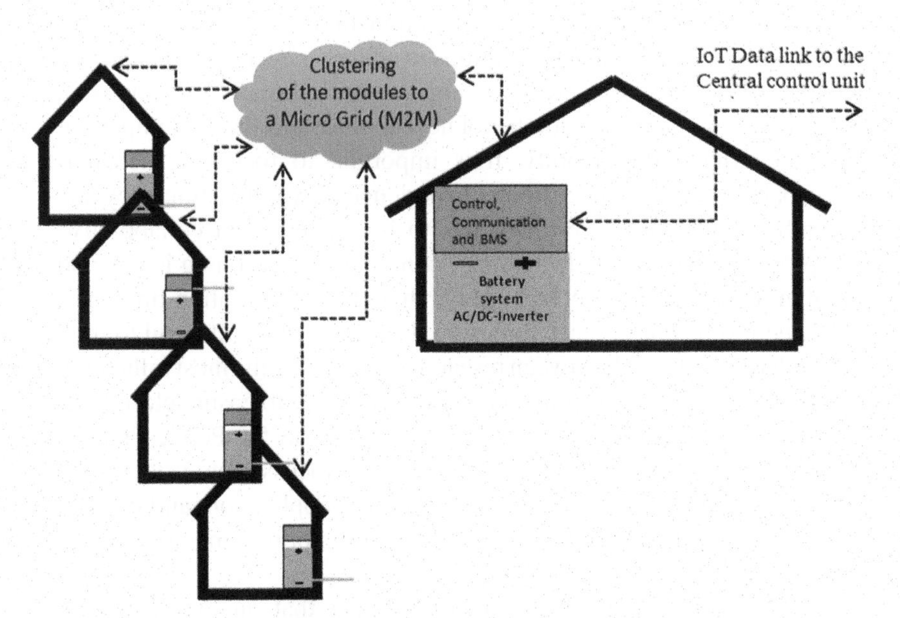

Figure 10.14 Clustering of the battery storage systems to organize itself.

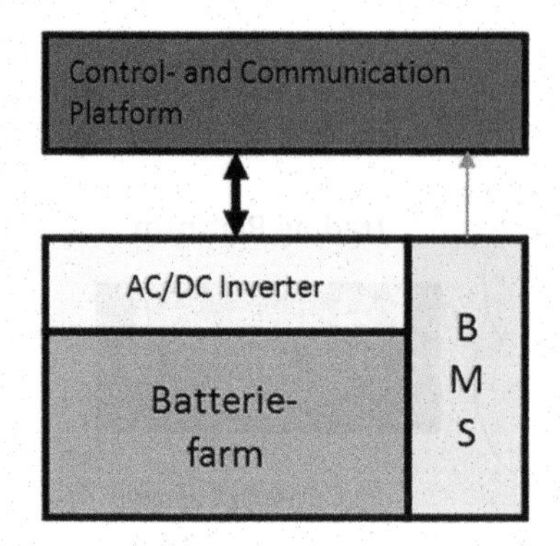

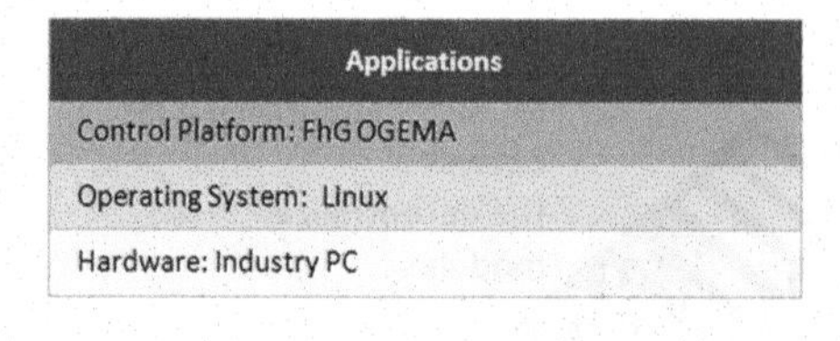

Figure 10.15 Control- and communication platform (IoT interface).

companies in Germany provide base to c. 100,000 buildings and c. 800,000 end users leading to 2,788 GWh power supply per year to end customers.

10.7 Conclusions and Outlook

The future communication standard 5G will change everyday life and economy dramatically. The success of a new generation of mobile communication system is mainly based on new applications to address new market potentials. 5G is expected to address industry, energy, mobility, real estate industry, health, etc. and generate a high number of new applications. Further important topics are security and reliability. It is important to focus on the areas network capability, consistent customer experience, flexibility, efficiency and innovation/8/. Significant investment has to be made in 5G collaborative research programs and within technology companies too. Timing is key and careful alignment is needed between research and standardisation timelines. Measures need to be put in place to ensure that research results lead to high quality standards. Close link between research projects and the most influential players in standardisation has proven to be the best way to ensure close link.

Within the EU research program Horizon 2020 the 5G Public-Private-Partnership (5G PPP) [12, 13] is established, which is focussed on systems with high data rate like video applications, systems for IoT applications and systems with low latency. Additional 5G will support automobile range, transport, manufacturing, banking, energy, smart cities, etc.

Finally, reference should be made to the fact that an excellent new generation will be successful only by learning from the mistakes of the past. Mistakes of the past were no terminals available or too late, less or wrong applications and too much focus on technology and not on business oriented applications. 3G and mainly 4G networks are already on a high level of performance and will be a high competition after 2020, too.

References

[1] George Calhoun, Digital Cellular Radio, Artech House, ISBN: 0-89006-266-8
[2] Ramjee Prasad, Werner Mohr, Walter Konhäuser, Third Generation Mobile Communication, Artech House, ISBN: 1-58053-082-6
[3] Walter Konhäuser: Keynote speech on IEEE Vehicular Technology Conference Fall 99, Amsterdam, September 19–22, 1999.
[4] An Introduction to LTE. 3GPP LTE Encyclopedia. Retrieved December 3, 2010.
[5] Long Term Evolution, a technical overview. Motorola. Retrieved July 3, 2010.
[6] News room press release. Itu. int. Retrieved 2012-10-28.
[7] ITU-R Confers IMT-Advanced (4G) Status to 3GPP LTE (Press release). 3GPP. 20 October 2010. Retrieved 18 May 2012.
[8] NGNM 5G White Paper – Executive Version, Version 1.0, 22nd December 2014.
[9] Walter Konhäuser, Smart Grid Applications as new Services within the Communication Networks, Global Wireless Summit 2014 in Aalborg from May 11–14, 2014.
[10] Smart Grids SRA 2035 Strategic Research Agenda, March 2012.
[11] OGEMA, http://www.iis.fraunhofer.de/de/ff/ener/proj/ogema-2-0.html
[12] Mohr. W., 5G Public-Private-Partnership in Horizon 2020 der EU-Commission, VDE dialog 02/2014, ITG news.
[13] EU Commission, Horizon 2020, https://ec.europa.eu/programmes/horizon2020/en/h2020-section/information-and-communication-technologies

About the Author

Walter Konhäuser was born in 1949. He studied electrical engineering at the Technical University in Berlin and finalized with Ph.D. In 2007 he was appointed to Professor for Mobile Communication Systems.

Since joining Siemens in 1982 he has been involved in a variety of assignments. For eighteen years he has worked in the mobile networks business and was CTO within the mobile infrastructure business of Siemens. In 2004 he became President for the WLAN business with worldwide responsibility.

After the merger of Siemens Communication Group with Nokia Net to Nokia Siemens Networks (NSN) in 2007 he became Head of Product Process Deployment and site manager for NSN in Berlin.

Since 2010 he works as independent professional as Executive Consultant and is partner since 2013 at Hoseit-Unternehmensberatung, Berlin and Cologne, where he is working on a decentralized energy storage concept based on Smart Grid applications and a centralized control system to contribute to the energy turnaround.

He is a member of the VDE, the German Association of Electrical Engineers and is Spokesman of the regional VDE organization in Berlin and Brandenburg.

11

5G: The Last Frontier?

Marina Ruggieri, Gianpaolo Sannino and Cosimo Stallo

University of Roma "Tor Vergata" – Center for Teleinfrastructures (CTIF_I), Italy

11.1 Introduction

The human body of wireless users at present plays an extremely limited role in both the network and the user terminal functions and architectures.

An innovative and challenging vision of the 5G-to-6G transition envisages the human body as a main actor with an active role in node and network functions as well as in user terminal architecture.

New frontiers in a tight connection of standardization and ethics (stethics) is intrinsically involved in the above vision.

In the last three years, the Center for Teleinfrastructures at the University of Roma "Tor Vergata" (CTIF_I) is focusing on a very challenging research effort where the maturity of Information and Communications Technology (ICT) is horizontally offered to deploy vertical applications in interdisciplinary areas, such as health, law, economics, cultural heritage, humanities, etc. In order to develop the above research approach, CTIF_I is composed of a strong group of ICT engineers, along with experts in the other disciplines, particularly in the clinical and biomedical areas.

The joint effort of the ICT and medical groups has produced an innovative and extremely challenging way of looking at the future of ICT, in particular at the meaning and the role of 5G in the evolutionary development of wireless systems.

In the above frame, the present Chapter, after highlighting the current achievements and open challenges of 5G, will present the CTIF_I vision on the challenges of the future beyond 5G world, providing an application example under investigation and early stage development with the dentistry component of the medical area. In the example, the human body – and in particular the oral cavity of the wireless user – plays an active role in the node of the networks.

The future work and perspectives of the CTIF_I approach and activities in the area of the human body active role to device and network functions of future mobile systems will be presented.

11.2 5G Achievements and Open Challenges

Mobile information systems have been facing the challenges of the continuously increasing demand for the high data rates and mobility required by new wireless applications. 4G (Fourth Generation)/LTE (Long Term Evolution) wireless networks now enable high-speed mobile web videos, IP (Internet Protocol) telephony, video gaming, mobile High-Definition (HD) TV, video conferencing, and even mobile 3D TV.

The current trend of annual doubling of wireless data traffic is expected to continue. Research on 5G wireless communications has started and developed rapidly. It is expected that 5G wireless infrastructure will be deployed beyond 2020. However, enabling technologies for 5G mobile information systems such as massive MIMO (Multiple Input Multiple Output), new modulation, and waveform design are still in their infancy. More efforts from this community are truly needed to make 5G mobile systems reach their full fruition.

Massive MIMO (M-MIMO) technologies have been proposed to scale up data rates reaching gigabits per second in the forthcoming 5G mobile communications systems. However, one of the crucial constraints is dimension in space to implement the M-MIMO.

How to make more small cells in sleeping state for energy saving in ultra-dense small cell system has become a research hotspot.

The key of wireless power transfer technology rests on finding the most suitable means to improve the efficiency of the system. The wireless power transfer system applied in implantable medical devices can reduce the physical and economic burden of patients since it will achieve charging in vitro.

Optimum beamforming and power allocation can be analyzed using, for instance, a game-theoretic framework [1] to maximize the rate of each user selfishly under the transmit power constraint and the Primary User (PU) interference constraint. The design of the cognitive MIMO system is formulated as a non-cooperative game, where the secondary users (SUs) compete with each other over the resources made available by the PUs.

Compressive sensing theory [2] can be applied to reconstruct the signal with far fewer measurements than what is usually considered necessary. While, in many scenarios, such as spectrum detection and modulation recognition, it is only expected to acquire useful characteristics rather than the original signals, where selecting the feature with sparsity becomes the main challenge.

With the widespread use of Internet, the scale of mobile data traffic grows explosively, which makes 5G networks in cellular networks become a growing concern. Recently, the ideas related future network, for example, Software Defined Networking (SDN) [3], Content-Centric Networking (CCN) [4], Big Data [5], have drawn more and more attention.

The exploitation of mm-waves in mobile cellular environments will represent a breakthrough of 5G, as clearly stated in [6]. In [6], the mobile urban propagation in the 38 GHz band is thoroughly investigated and the viability of broadband transmission by Line Of Sight (LOS) and Not Line of Sight (NLOS) links is demonstrated. In [7], the wireless communication channel operating in the 70 GHz and 80 GHz band is characterized to define specific parameters and constraints that can be used to define the optimal topology of a multi-gigabit access network. P-t-P links with 10 Gb/s capacity are considered. In [8], single-carrier higher-order modulations and FDM (Frequency Division Multiplexing) channel aggregation techniques are used in combination with MIMO spatial multiplexing for P-t-P multi-gigabit short-distance E-band LOS connections, pushing the available data-rate up to 48 Gb/s. The use of E-band for P-t-P backhauling is a possible scenario.

Well-known backhaul technologies such as spectral-efficient LOS microwave, fiber and copper are being tailored to support the exploding user traffic demand in the RAN (Radio Access Network). However, owing to their position below roof height, a substantial number of small cells in urban settings do not have access to a wired backhaul, or clear line of sight to either a macro cell or a remote fiber backhaul point of presence. For distances of the order of 1-1.5 kilometers, which are expected to be enough for Long Term Evolution (LTE)-A small-cell backhauling, mm-wave systems can provide backhauling in the order of several gigabits per second.

Figure 11.1 shows a potential multi-tier LTE-A cell deployment with mm-wave wireless backhaul. Small-cell Base Transceiver Station (BTS) are installed closer to smaller buildings or over streetlamps. Macro BTS are installed on the top of the bigger buildings. The presence of LOS connection is guaranteed for P-t-P backhaul among macro BTSs and may be ensured for P-t-P among small cell BTSs installed on the streetlamps. Also P-t-mP backhaul among macro BTSs and small-cell BTSs on the streetlamps may be

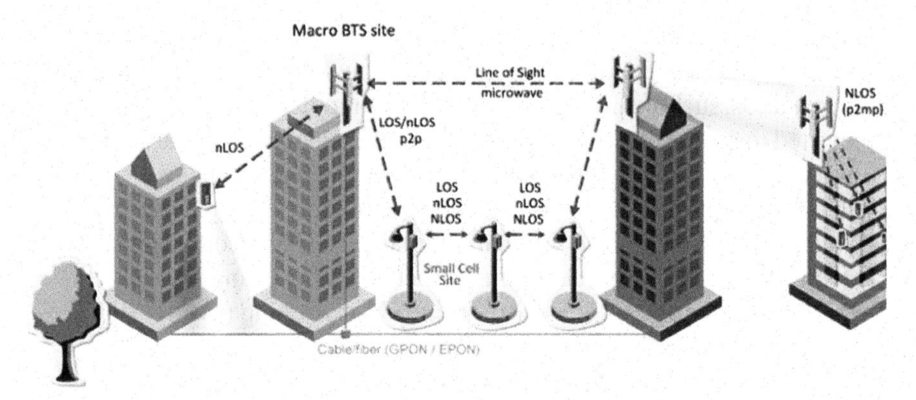

Figure 11.1 Typical configurations of small-cell backhaul in LTE-A standard.

reasonably assured. These LOS links can be effectively supported by E-band multi-gigabit transmission.

In [9] E-band for P-t-P backhauling in LTE-A small cells is considered in combination with point-to-multipoint NLOS backhauling at 6 GHz. The solutions envisaged in [9] looks rather similar to the commercial products for small-cell backhauling supplied by CERAGON corporation (Palo Alto, CA) [10]: the FibeAir-70^{TM} for LOS P-t-P transmitting at aggregate data-rate of 1 Gb/s at 81 GHz, and the FibeAir-2500^{TM} for NLOS P-t-mP transmitting at 200 Mb/s at 6 GHz. All these components rely on single-carrier transmission techniques.

State-of-the-art PHY-layer solutions considered in [7–9] look dated and lack of efficiency in particular in the presence of phase-noise, nonlinear distortions and signal-to-noise ratio drops due to atmospheric attenuation and rain fading. Such kind of impairments is usual in mm-wave transmission. Therefore, alternative solutions should be investigated in order to allow robust small-cell backhaul with reduced power expense.

A completely different typology of transmission based on UWB (Ultra-Wide Band) techniques has been considered in [11, 12]. However, as shown in [13] and [14], IR (Impulse Radio) techniques, thanks to their simplicity and intrinsic robustness, are very efficient also if employed in the E-band context. The IR-based backhaul relies on Pulse-Position-Modulation (PPM) for digital transmission and on Time Hopping (TH) for multi-user P-t-mP transmission. Such transceiver architecture is very efficient, very simple, and cost-effective and can be implemented by using Commercial Off the Shelf (COTS) hardware.

Finally, the 5G paradigm is paving the way for rich healthcare services enhanced with video calls and high definition images, available nearly everywhere. 5G networks are supporting sensors requiring very low energy consumption, with one battery charge every fifteen years. It will be very beneficial for medical connected devices such as blood pressure or insulin body worn sensors.

Furthermore, the high reliability and security of 5G infrastructures should help to alleviate the legitimate end user and health professionals concerns about privacy and hacking around health data and services. When thinking about new 5G capabilities in the domain of lower latency, reaching the target of 1 ms, new applications in the area of prosthesis or augmented human can be conceived.

The above frame about 5G suggests that, in spite of the open challenges and some potential issues, the 5G system paradigm is solid and it could represent a (final?) landing point for the mobile world as we have been used to know it from the introduction of GSM [15].

The 5G approach, where the extreme efficiency and speed of the mobile network as well as the full convergence between wired and wireless are the ultimate goals, indicates that 5G could in principle "saturate" the achievable maturity at system level from both operators and users viewpoints.

It is therefore very natural to pose the question of what could happen to the networks and users after the full deployment of 5G.

A possible answer is provided in the rest of the Chapter.

11.3 The Challenges on the 5G Frontier

The transition between 5G and a following generation can introduce dramatic changes in the role played by the user inside the network.

A medium-term scenario envisages an increasing synergy between the user device capabilities and their implementation through the user's body. The result will be a highly personalized body-related device that will be able to solve intrinsic limitations in the current interaction with the device (e.g., finger size w.r.t keyboard).

A further step ahead allows predicting a strong interaction between the brain functions of the user and the device. A possible scenario would envisage the device be driven by brain and emotional related commands. In the depicted frame, the device would need not only for pure technology but also for a proper merging between standardization and ethics (stethics).

In parallel, the biomedicine is progressing dramatically, and the early diagnostics of important diseases can take advantage of proper chips to be implanted in the body. A label-free spiral microfluidic device is proposed, for instance, to allow size-based isolation of viable Circulating tumor cells (CTCs), i.e. rare cancer cells that are shed from primary or metastatic tumors into the peripheral blood circulation. Phenotypic and genetic characterization of these rare cells is crucial to provide important information and guide cancer staging and treatment (e.g., [16, 17]).

The increased closeness between the human body and functions that are generally assigned to an external device is the starting point of a research effort played by a joint team of engineers and medical doctors to identify a set of functions that can be supported by one or more chips to be implanted. Those chips can be supporting not only "pure medical" functions but also – and here is the matter that can be of interest for the post 5G community – device and network functions.

This is the key-passage to achieve a partially internal user device (Body as a Device – ByD) and to assign to each user functions related to the nodes of the network (Body as a Node – ByN).

A pictorial view of the above concept is shown in Figure 11.2.

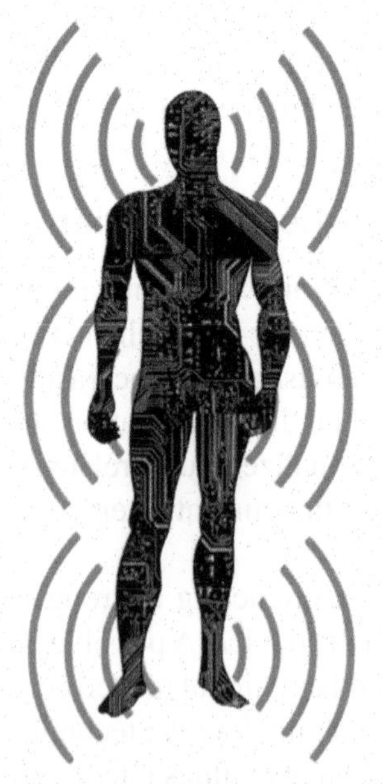

Figure 11.2 Pictorial view of the body as a device and as a node concept.

In the depicted frame, the next paragraph describes a possible scenario where the ByD and/or ByN approaches could be implemented by implants in the oral cavity.

11.4 The Oral Cavity as Device or Node

The oral cavity is one of the main gateways to the body. It is in fact in the early months of life, a cognitive tool which subsequently develops with all its structures (mucous membranes, muscles, tendons, bones, joints, teeth, nerves, vessels, glands, etc) as a body involved in carrying out different functions (immuno-protection, digestion, chewing, swallowing, phonetic). The complexity as well as the sensitivity of these structures, makes oral health closely related to general health conditions. The first evidences of many systemic diseases are borne by the oral structures and therefore are diagnosed through an examination of those.

Precisely for this reason, the oral cavity may constitute the ideal environment for positioning sensory systems and collecting data that might be useful both in monitoring oral activities, detecting pathologies/problems and planning early intervention and partial remote management.

In addition, as highlighted in Section 11.3, the devices could also host functions of the user device (Oral Cavity as a Device – OCD) and/or of the network node (Oral Cavity as a Node – OCN).

In Figure 11.3 possible positioning of sensors and chips in the oral cavity is depicted.

Due to the ICT contribution performance and costs of cures for both the public health system and the private patients could change, especially in the Dentistry field, where a significant growth of technology-driven approaches is occurring. In recent years technological advancement have radically changed the face of dentistry by increasing the reliability of diagnostic imaging, simplifying treatments planning/realization thanks to computer-guided and minimally invasive approaches, reducing working time and errors human-related thanks to the intraoral optical scans and CAM (Computer Aided Manufacturing) units [18–20]. However the potential opportunities offered by this sector have not been truly exploited yet, since these technological advancements have focused on aspects of close dental relevance.

Using intraoral micro-biosensors featuring a wireless transmission, instead of wired connections as proposed in the past two decades by several researchers [21–27], could open new scenarios for dentistry applications (related to oral health), general sensing applications (sensing parameters

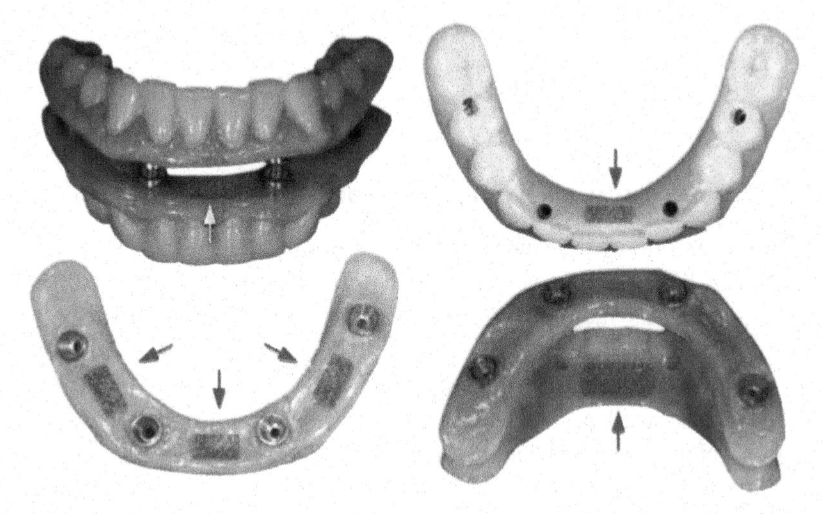

Figure 11.3 Possible location of sensors and chips for the OCD or OCN approach.

related to human health) as well as Human-Machine-Interface (HMI) applications (using measurements of oral cavity to drive/control machines). The advancements in micromachining techniques have allowed sensor miniaturization for minimum discomfort in the oral cavity and easy implantation with no tissue injury. The sensors can be easily integrated into prosthetic devices and orthodontic appliances, even in the teeth or bones. [28] Chemical markers relevant to health may be detected by electrochemical sensors, as well as variation in ph, temperature, pressure or position of the oral environment.

The use of intraoral sensors from the communication point of view is analogous to other mHealth applications. However wireless communication between intraoral devices and the outside world is challenging, since the oral cavity is a hybrid propagation environment. Intraoral sensors are in the border between intra-body and wearable sensors. It is not strictly speaking a communication "inside the body" as a sensor inside the body, depending on its anatomical location, is usually surrounded by a distinct and fairly stable tissue environment.

An intraoral device is located in a constantly changing environment depending on the relative positions of the jaws and movements of the tongue, which continuously changes shape when one swallows, breathes, or speaks. As intraoral devices are in contact with the gums, the tongue, or the palate, and their associated receivers may be in contact with the human body, Body Channel Communication (BCC) can be considered to be used for these devices [29]. However, when applied to intraoral devices, neither of the BCC methods maintains their desired operating conditions due to attenuation effect. Therefore, the use of RF-based wireless communications could be considered. As a matter of fact, RF wireless communication has been already successfully adopted in implantable devices such as the pacemaker and neuro-stimulator.

The proper selection of the frequency band is fundamental for establishing a reliable link between intraoral devices and the outside world. The selection of the frequency is strictly related to two fundamental design elements: size of the antenna and attenuation. It is worth noted that an RF-based communication approach foresees the use of a battery inside the mouth, at least for those important applications that foresee a continuous monitoring. This is one of the main challenges limiting the adoption of intraoral devices for continuous monitoring. Interesting solutions for recharging the sensors are based on the use of the jaw movement that normally occurs when chewing, eating and speaking [30]. For instance, one can obtain approximately 580 J only from daily chewing, which is equivalent to an average power of approximately 7 mW. An alternative approach has been proposed in two studies [31, 32] where a passive sensor is inserted in the mouth and read by an external device.

It should be highlighted that the presence of the sensors inside the oral cavity might cause metabolic alterations (causing inflammation or tissue damage). Therefore, the molecular and biochemical approaches undertaken to unravel possible metabolic alterations (with particular reference to quantification of recognized markers of inflammation or tissue damage), occurring during prosthesis implantation and long-term wearing, should be addressed [33].

The presence of sensors and chips has to be, therefore, analyzed also in terms of potential harm to the oral cavity, in particular, and the whole body in general. The sensor/chip design and implementation have to be driven by the guidelines and requirements deriving from the above analysis. The same applies to whatever part of the body is selected for the chip positioning.

11.5 Future Work

The Chapter has provided a possible vision of post 5G systems, particularly related with the active use of the human body of the user to both extend, integrate or substitute the user device functions and/or the node functions.

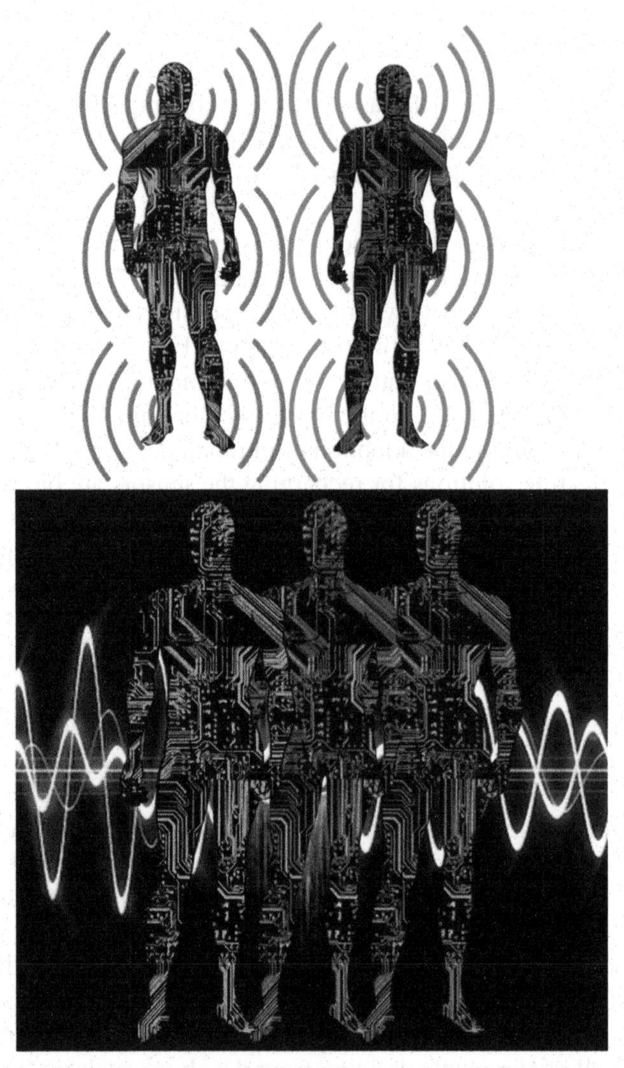

Figure 11.4 Possible multi-user, multi-device and multi-node scenario.

In the envisaged vision, the scenarios reported in Figure 11.4 are possible. A number of important activities are being and will be undertaken to understand fully the impact and the potential of the ByD and ByN approaches, and to design and implement harmless chips. The experimental activity will start from the OCD and OCN cases and will be later extended to other parts of the body.

References

[1] Feng Zhao, Hongsheng Wu, Hongbin Chen, and Wen Wang, "Game-Theoretic Beamforming and Power Allocation in MIMO Cognitive Radio Systems with Transmitter Antenna Correlation", Mobile Information Systems, Hindawi Journal, Volume 2015 (2015).

[2] B. Adcock and A. C. Hansen. Generalized sampling and infinite dimensional compressed sensing. Preprint, 2012.

[3] Hyojoon Kim and Nick Feamster, "Improving Network Management with Software Defined Networking", IEEE Communication Magazine 2013.

[4] Diego Perino and Matteo Varvello "A Reality Check for Content Centric Networking", ICN'11, 2011, Toronto, Ontario, Canada.

[5] Christopher J. Bucholtz. (2012, March) Customers, Big Data, and the Internet of Things. [Online]. http://www.technewsworld.com/story/74692.html

[6] T. Rappaport, S. Sun, R. Mayzus, H. Zhao, Y. Azar, K. Wang, G. N. Wong, J. K. Schulz, M. Samimi and F. Gutierrez, "Millimeter-wave Communications for 5G Cellular: It Will Work!," IEEE Access, vol.1, 2013, pp. 335–349.

[7] C. Colombo, and M. Cirigliano, "Next-Generation Access Network: A Wireless Network Using E-Band Radio Frequency (71–86 GHz) to Provide Wideband Connectivity," Bell Labs Tech. Jour, vol.16, no.1, 2011, pp. 187–206.

[8] C. Sacchi, C. Stallo, and T. Rossi, "Space and Frequency Multiplexing for MM-Wave Multi-Gigabit P-t-P Transmission Links," Proc. of 2013 IEEE Aerospace Conf., Big Sky (MT), March 2013, pp. 1–10.

[9] J. Zhang, W. Ni, J. Matthews. C-K. Sung, X. Wang, H. Suzuki, and I. Collings, "Low Latency Integrated Point-to-Multipoint and E-band Point-to-Point Backhaul for Mobile Small Cells," Proc. of IEEE ICC 2014 Workshops, June 2014, pp. 592–597.

[10] CERAGON: Wireless backhaul solutions for small cells. Available: http://www.ceragon.com/files/Ceragon

[11] C. Stallo, S. Mukherjee, E. Cianca, M. Ruggieri, "IR-UWB Approach for High Bit Rate Transmission in the E Band," Proc. of IEEE ESTEL Conf., Oct. 2012, pp. 1–6.

[12] C. Stallo, S. Mukherjee, E. Cianca, M. De Sanctis, T. Rossi, M. Lucente, M. Ruggieri, "UWB for high bit rate communications beyond 60 GHz", Proc. of IEEE Int. Conf. on Personal Indoor Mobile Communications (PIRMC) 2010, Istanbul, Turkey, May 2010.

[13] Rahman, T. F.; Sacchi, C.; Stallo, C., MM-wave LTE-A small-cell wireless backhauling based on TH-IR techniques, IEEE Aerospace Conference 2015, Big Sky, Montana, Pages 1–9.
[14] C. Stallo, E. Cianca, S. Mukherjee, T. Rossi, M. De Sanctis, M. Ruggieri, "IR-UWB for multi-gigabit communications beyond 60 GHz", Telecommunication Systems Journal, Springer US, 2014.
[15] Prasad R., Ruggieri M., (2003) "Technology Trends in Wireless Communications", Artech House, Boston, 2003, ISBN 1-58053-352-3.
[16] Warkiani ME, Khoo BL, Wu L, Tay AK, Bhagat AA, Han J, Lim CT. Ultra-fast, label-free isolation of circulating tumor cells from blood using spiral microfluidics. Nat Protoc. 2016 Jan; 11(1): 134–48.
[17] Warkiani ME, Guan G, Luan KB, Lee WC, Bhagat AA, Chaudhuri PK, Tan DS, Lim WT, Lee SC, Chen PC, Lim CT, Han J. Slanted spiral microfluidics for the ultra-fast, label-free isolation of circulating tumor cells. Lab Chip. 2014 Jan 7; 14(1): 128–37.
[18] Sannino G, Barlattani A. Straight versus angulated abutments on tilted implants in immediate fixed rehabilitation of the edentulous mandible: a 3-year retrospective comparative study. Int J Prosthodont. 2016;29(3) doi: 10.11607/ijp.4448
[19] Sannino G. All-on-4 concept: a 3-dimensional finite element analysis. J Oral Implantol. 2015 Apr; 41(2): 163–71.
[20] Sannino G., Germano F., Arcuri L., Bigelli E., Arcuri C., Barlattani A. CEREC CAD/CAM Chairside System. Oral Implantol (Rome). 2015 Apr 13; 7(3): 57–70.
[21] K. Hori, T. Ono, K. Tamine, J. Kondo, S. Hamanaka, Y. Maeda, J. Dong, M. Hatsuda, "Newly developed sensor sheet for measuring tongue pressure during swallowing", J. Prosthod. Res. 53, pp. 28–32, 2009.
[22] K. Nishigawa, E. Bando, M. Nakano, "Quantitative study of bite force during sleep associated bruxism", J. Oral Rehab. 28, pp. 485–491, 2001.
[23] Q. Peng, T. F. Budinger, "ZigBee-based wireless intra-oral control system for quadriplegic patients", IEEE Transactions on Engineering in Medicine and Biology Society, pp. 1647–1650, 2007.
[24] E. Sardini, M. Serpelloni, R. Fiorentini, "Wireless intraoral sensor for the physiological monitoring of tongue pressure", IEEE Transactions on Solid-State Sensors, Actuators and Microsystems, pp. 1282–1285, 2013.
[25] H. Park, J. Kim, M. Ghovanloo, "Development and preliminary evaluation of an intraoral tongue drive system", IEEE Transactions on Engineering in Medicine and Biology Society, pp. 1157–1160, 2012.

[26] H. Park, M. Kiani, H. Lee, J. Kim, J. Block, B. Gosselin, M. Ghovanloo, "A wireless magnetoresistive sensing system for an intraoral tongue-computer interface", IEEE Transactions on Biomedical Circuits and Systems, pp. 571–585, 2012.

[27] J. H. Kim, J. H. Jung, A. Y. Jeon, S. H. Yoon, J. M. Son, S. Y. Ye, G. R. Jeon, "System development of indwelling wireless pH Telemetry of intraoral acidity", IEEE Transactions on Information Technology Applications in Biomedicine, pp. 302–305, 2007.

[28] Sannino G., Cianca E., Hamitouche C., Ruggieri M. M2M Communications in Dentistry: a wireless communications perspectives. Chapter Book title: Future Access Enablers for Ubiquitous and Intelligent Infrastructures Volume 159 of the series Lecture Notes of the Institute for Computer Sciences, Social Informatics and Telecommunications Engineering pp 118–124 First International Conference, FABULOUS 2015, Ohrid, Republic of Macedonia, September 23–25, 2015. DOI 10.1007/978-3-319-27072-2_15

[29] Cho, N., Yoo, J., Song, S., Lee, J., Jeon, S., Yoo, H. The human body characteristics as a signal transmission medium for intrabody communication. IEEE Trans. on Microwave Theory and Techniques, 1080–1086, (2007).

[30] Delnavaz, A., Voix, J. Flexible piezoelectric energy harvesting from jaw movements. Smart Mater. Struct. 23, 1–8 (2014).

[31] Mannoor, M. S., et al., Graphene-based wireless bacteria detection on tooth enamel. Nat. Commun. 27, 763 (2012).

[32] Diaz Lantada, A., González Bris, C., Lafont Morgado, P., Sanz Maudes, J.: Novel System for Bite-Force Sensing and Monitoring Based on Magnetic Near Field Communication, Sensors, 12, 11544–11558 (2012).

[33] Sannino G., Cianca E., Coletta M., Prasad R., Ruggieri M., Sbardella D. Integrated Wireless and Sensing Technology? for Dentistry? an Early Warning System for Implant-Supported Prosthesis. 2015 IEEE International Symposium on Systems Engineering (ISSE), Rome September 28–30.

About the Authors

Marina Ruggieri is Full Professor of Telecommunications Engineering at the University of Roma "Tor Vergata" and therein member of the Board of Directors.

She is IEEE 2016 Vice President-Elect, Technical Activities. She is Past Director of IEEE Division IX (2014–2015).

She is member of the IEEE Public Visibility and Fellow Committees. She has been member of the IEEE Governance Committee. She is Sr. past President of the IEEE Aerospace and Electronic Systems Society.

She is proboviro of the Italian Industries Federation for Aerospace, Defense and Security (AIAD); member of the Technical-Scientific Committee of the Center for Aeronautical Military Studies. She has been Vice President of the Roma Chapter of AFCEA (2006–2015).

She is co-founder and Chair of the Steering Board of the interdisciplinary Center for Teleinfrastructures (CTIF) at the University of Roma "Tor Vergata". The Center, that belongs to the CTIF global network, with nodes in USA, Europe and Asia, focuses on the use of the Information and Communications Technology (ICT) for vertical applications (health, energy, cultural heritage, economics, law) by integrating terrestrial, air and space communications, computing, positioning and sensing.

She is Principal Investigator of the 40/50 GHz TPD#5 Communications Experiment on board Alphasat (launched on July 2013).

She received: 1990 Piero Fanti International Prize; 2009 Pisa Donna Award as women in engineering; 2013 Excellent Women in Roma Award; Excellent and Best Paper Awards at international conferences.

She is IEEE Fellow. She is author/co-author of 330 papers, 1 patent and 12 books.

Gianpaolo Sannino is a CTIF Global Capsule (CGC) funder member and graduated in dentistry from the University of Rome Tor Vergata (URTV, Rome, Italy). He obtained a first Ph.D. in Odontostomatological Sciences and a second Ph.D. in Materials for Health Environments and Energy, both at the URTV. He has been Adjunct Professor at the URTV teaching General Dentistry and Oral Implantology. He has further been Adjunct Professor at UZKM (Tirana, Albania) teaching General Dentistry, Prosthodontics and Gnathology. He is now Adjunct Professor at the Postgraduate School in Hospital pharmacy (Pharmacy School) teaching Pain Management at University of Salerno (Salerno, Italy). He is also clinical and didactic tutor at Vita-Salute University, San Raffele Hospital (Milan, Italy). He participated to several European and national projects. His background is on: Prosthodontics, Periodontology and Oral Implantology (bone and tissue regeneration procedures), endodontics, radiology, oral surgery, material sciences, biomechanics of dental system, numerical methods. His recent research interest has moved towards Information and Communications Technology (ICT) related to Health services. He is the vice-chair Chair of the Steering Board of the interdisciplinary Center for Teleinfrastructures (CTIF) at the URTV. He is author of several papers, on international journals/transactions and proceedings of international conferences.

Cosimo Stallo graduated in Electronic Engineering with specialization in Microelectronic Design at Polytechnic of Bari in 2005. He was involved in studying, analysing and simulating Optical Model of a selectively oxidized VCSEL (Vertical Cavity Surface Emitting Laser) emitting at 850 nm. He received cum laude a M.Sc. Degree in "Advanced Communication and Navigation Satellite Systems" in 2006 at University of Rome Tor Vergata. He obtained a scholarship funded by Italian Space Agency for the Ph.D. course in Microelectronics and Telecommunications XXII cycle (2006–2009) at University of Rome "Tor Vergata". He obtained cum laude the Ph.D. in Microelectronics and Telecommunications at Electronic Department of Engineering Faculty of University of Rome "Tor Vergata" on June 3 2010 with the Thesis: "Wireless Technologies for Future Multi-Gigabit Communications Beyond 60 GHz: Design Issue and Performance Analysis for Terrestrial and Satellite Applications".

In 2007 he entered in Center for TeleInFrastruktur (CTIF), an international research centre, operating in the fields of ICT, with offices in Denmark (Aalborg) (headquarter), Italy (Rome), India, Japan and U.S. (www.ctif.org).

Currently, he is the Assistant Professor for the course in Signal Processing and Power Electronics at Mechatronic Engineering, University of Rome Tor Vergata in Colleferro (Rome). Since 2011 he is Professor for the course on Satellite Navigation at Master of Science on "Advanced Satellite Communication and Navigation Systems" of University of Rome Tor Vergata, Rome. Since 2013–14 he is tutor for the course on Communication Systems at University of Trento, Italy.

From May 2007 to January 2015 he was the Italian Chair of the IEEE Professional Activities Affinity Group (formerly GOLD).

Since February 2010 he is the Chair of the Space Systems Technical Panel. From 2010 to 2015 he was the Chair of the IEEE AESS Professional Activities Affinity Group.

He was the Editor in Chief of the IEEE AESS Quarterly Email Blast (QEB) since October 2011 from December 2013. Since 2013 he participates to the initiative IEEE Division IX for Quality of Life (Div IX-4-QoL). Since 2014 he is involved in TAB Awards and Recognition Committee (TABARC). His main fields of research concern terrestrial and satellite communications at Extremely High Frequency (EHF), design of payloads and subsystems for telecommunications at Q/V [35–75 GHz] and W [75–110 GHz] bands, satellite navigation, signal processing, and ICT for Quality of Life (biotechnology, health and energy). He is co-author of about 50 papers on international journals/transactions and proceedings of international conferences.

12

802.11ax for 5G

Richard van Nee

Qualcomm, the Netherlands

12.1 Introduction

It is expected that 5G is made up of several radio access technologies as there is probably no single technology capable of meeting all requirements. One likely candidate for being part of 5G is the new 802.11ax standard. This wireless LAN standard offers data rates up to 10 Gbps in channels up to 160 MHz. It improves capacity for large number of users by employing OFDMA which reduces overhead relative to the TDD approach of existing wireless LAN standards. At the same time, the use of uplink OFDMA provides a significant range benefit for clients that have less transmit power than access points, a situation that is pretty common for handheld devices. Further capacity enhancements are provided by the use of uplink MU-MIMO in addition to downlink MU-MIMO that was already present in the 802.11ac standard.

Peak throughputs of wireless LAN have grown exponentially for quite some time as shown in Figure 12.1. From 2 Mbps for early 802.11 products in the nineties, peak rates have steadily climbed to several Gbps for current 802.11ac products and expect to reach approximately 10 Gbps with 802.11ax [1]. While early 802.11 standards primarily focused on single link rates, the attention shifted towards network throughput since 802.11ac with the introduction of MU-MIMO [2]. Improving network throughput in dense networks is the primary goal of 802.11ax. This is important to deal with increased levels of inter-network interference and to provide more capacity for networks with large numbers of clients, for instance in cases where WiFi is used for data offload to relieve the pressure on LTE spectrum. The latter use case is one of the main drivers to make WiFi part of 5G.

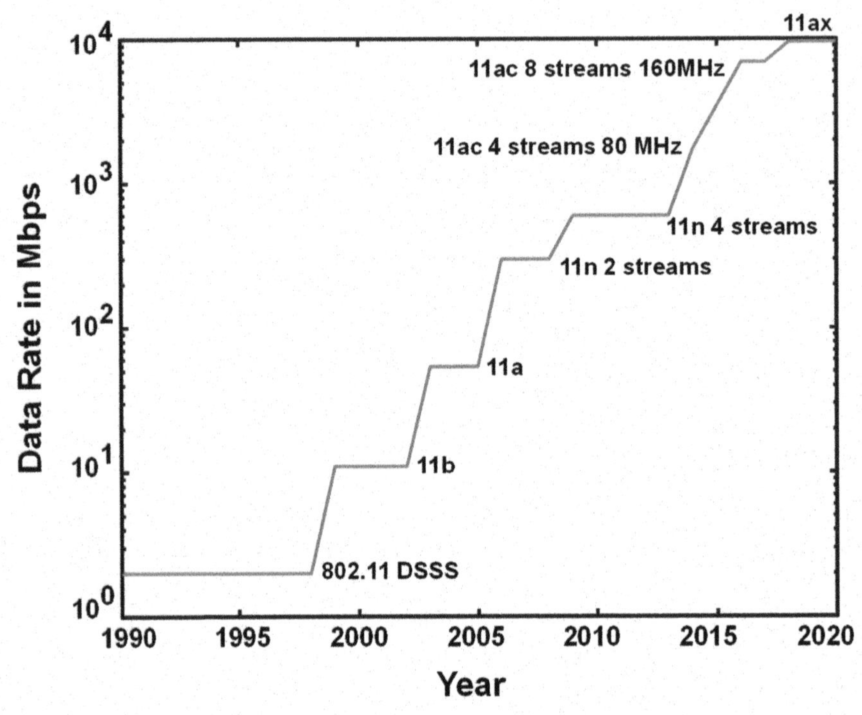

Figure 12.1 WLAN data rate growth.

12.2 802.11ax Features

The main new features in 802.11ax are:

1. Longer symbol durations with 4 times smaller subcarrier spacing
2. OFDMA
3. Uplink MU-MIMO
4. 1024-QAM
5. Range extension with DCM
6. Dynamic CCA

Longer symbols with 4 times smaller subcarrier spacing are used for several reasons. First, it provides a larger data rate by reducing the relative duration of the guard interval and by a slight increase of the band filling. Second, it allows the use of larger guard intervals which provides more delay spread robustness. Guard intervals of 800 ns, 1600 ns and 3200 ns are defined in 802.11ax together with a 12.8 microseconds FFT interval whereas 802.11ac uses a guard interval of 800 ns or 400 ns with a 3.2 microseconds FFT interval.

The additional delay spread robustness is especially useful for outdoor use where RMS delay spreads in the order of a microsecond may occur. A third reason for using longer symbol durations is that it provides a more relaxed synchronization requirement for uplink OFDMA and uplink MU-MIMO.

To enable channel estimation, an 11ax preamble contains a number of HE-LTF symbols equal to or larger than the total number of spatial steams, similar to 11ac. One disadvantage of using larger symbol durations is that the overhead of this channel estimation would be 4 times larger than for 11ac if all HE-LTF symbols used the longest symbol interval of 16 microseconds. To reduce this overhead, 11ax specifies 4 different modes of operation with symbol durations of 4 μs (including 800 ns guard interval), 7.2 μs (including 800 ns guard interval), 8 μs (including 1.6 μs guard interval), and 16 μs (including 3.2 μs guard interval). All modes beside the longest 16 μs require tone interpolation to get the channel estimates for all data tones.

For non-OFDMA mode, 11ax uses 242 tones in 20 MHz including 8 pilots, giving a peak data rate of 1147 Mbps for 8 streams, 1024-QAM, coding rate 5/6, with 800 ns guard interval. For 40 MHz, 484 tones are used including 16 pilots, giving a peak rate of 2294 Mbps. 80 MHz uses 996 tones with 16 pilots, which are also used by 160 or 80 + 80 mode in each 80 MHz segment, giving a top rate for 80+80 of 9607.8 Mbps for 8 spatial streams and 1201 Mbps for a single spatial stream.

12.3 Interoperability and Mode Detection

The new 802.11ax standard will be used both in the 2.4 GHz and 5 GHz band. Devices that implement 802.11ax have to be interoperable with existing 802.11 products which mean they need to support all existing 802.11 waveforms for either the 2.4 or 5 GHz band. At the same time, the new 802.11ax waveforms should not cause performance degradation for existing 802.11 devices that did not implement 802.11ax. To achieve the latter, all 802.11ax waveforms start with a 802.11a-like preamble, similar to 802.11n and 802.11ac. The signaled rate in the 802.11a preamble is set to 6 Mbps and the length field is set to match the duration of the 802.11ax packet. This ensures that all devices that can receive the 802.11a part will properly defer for the entire packet even though those devices may not be able to receive any 802.11ax data.

The existing preamble structures for 802.11a, 11n, and 11ac are depicted in Figure 12.2. It is important that a device can quickly and reliably detect what preamble type is being received. For 11a, 11an and 11ac, this detection

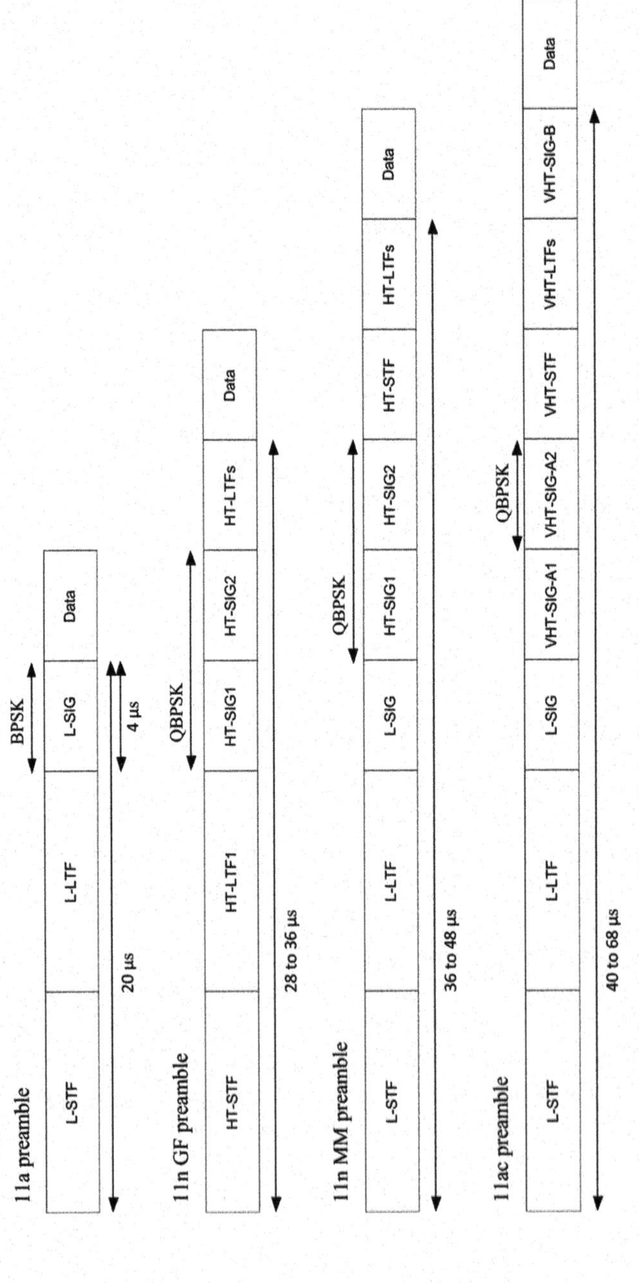

Figure 12.2 Preamble structures for 11a, 11n and 11ac.

is based on the use of QBPSK at different parts of the preamble. QBPSK is equal to BPSK rotated by 90 degrees. An 11n greenfield preamble uses QBPSK at the first symbol after the LTF symbol. An 11a packet has a BPSK L-SIG symbol at this same location. Hence, a receiver can detect whether a packet is 11a or 11n-GF by estimating whether the constellation of this first symbol after LTF is BPSK or QBPSK. An 11n mixed-mode (MM) preamble has a BPSK L-SIG followed by two QBPSK HT-SIG symbols. The L-SIG symbol is a valid 11a-type signal field indicating the use of 6 Mps BPSK and with a length field that covers the entire 11n-MM packet. After decoding this L-SIG symbol, a receiver does not know yet whether the packet is 11a or 11n-MM, it only knows it cannot be 11n-GF based on the QBPSK check on L-SIG indicating BPSK. By doing a second QBPSK check on the symbol after L-SIG, it can detect whether the packet is 11n-MM versus 11a. For 11ac packets, a third QBPSK check is required at the second symbol after L-SIG. The packet is detected as 11ac based on the L-SIG being BPSK and signaling 6 Mbps, the first symbol after L-SIG also being BPSK, and the second symbol after L-SIG being QBPSK. Devices that did not implement 11ac will detect the 11ac packet as 11a based on the valid L-SIG symbol and the absence of QBPSK in the first two symbols after the LTF symbol.

Figure 12.3 shows the additional preambles defined by 11ax. There are 4 new preambles for Single User (SU) mode, Multi User (MU) modes being downlink MU-MIMO and downlink OFDMA, Trigger modes being uplink MU-MIMO and uplink OFDMA, and extended range mode. For existing devices that did not implement 11ax, all 11ax packet types are detected as 11a such that a proper defer will be done based on the L-SIG length field that is set to cover the entire duration of the 11ax packet. An 11ax device can discriminate 11ax versus 11n and 11ac based on the fact that the L-SIG length field modulo 3 is 1 or 2 for 11ax versus 0 for 11n and 11ac. The L-SIG length field indicates a number of bytes, but 11n, 11ac and 11ax only use this length field to indicate the total duration of the packet and not the byte length. Since the spoofed 11a rate of 6 Mbps gives 3 bytes per symbol, the same packet duration can be signaled by 3 different L-SIG length values with 1, 2, or 3 bytes in the last symbol. If the L-SIG length modulo 3 is 1 or 2, then a receiver knows that the packet cannot be 11n or 11ac, but it does not know yet whether it is 11a or 11ax. To detect this, there are 2 different checks that can be made. First, the L-SIG is repeated in 11a packets. This repetition can be used for mode detection. At the same time, it provides a range advantage in the extended range 11ax mode where all fields use half the lowest 11a rate. Second, the L-SIG, repeated L-SIG, HE-SIG-A,

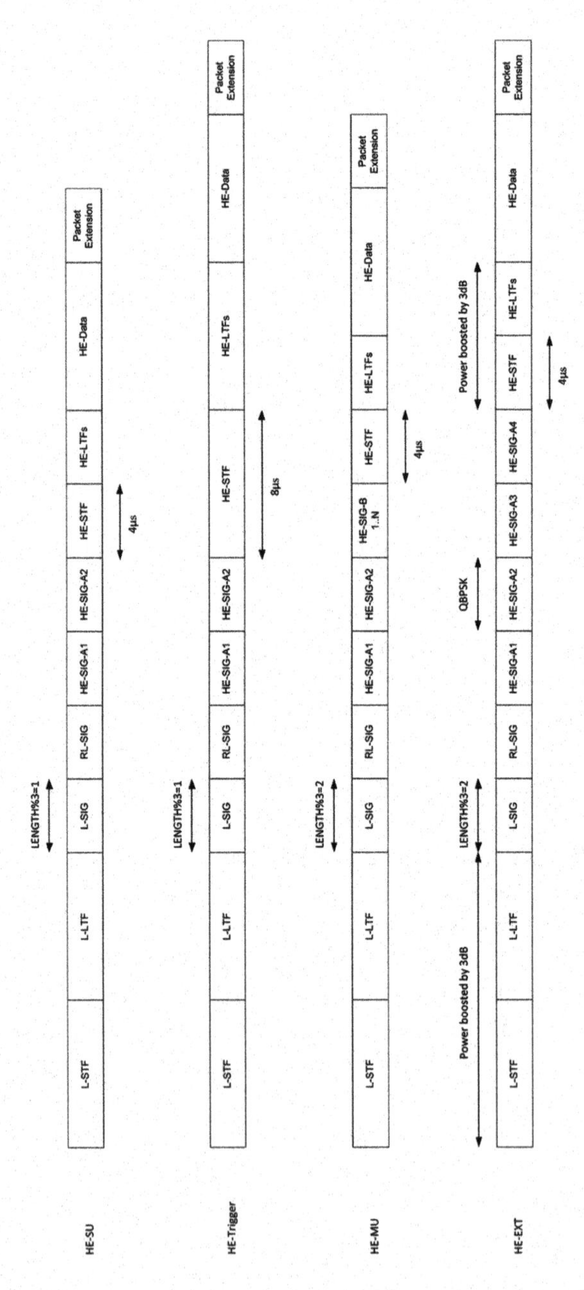

Figure 12.3 Preamble structures for 11ax.

and HE-SIG-B symbols all have 4 edge tones more than 11a symbols. These extra tones contain known BPSK values and they can be used both for mode detection as well as for improving channel estimation on the HE-data.

Once a receiver detected that a packet is 11ax and finds out the specific 11ax packet type, it starts to decode signaling data in HE-SIG-A. This field contains data that is needed to decode the rest of the packet such as the MCS (BPSK up to 1024-QAM, coding rate ½ up to 5/6), number of spatial streams, guard interval and LTF size, bandwidth, SU/MU, STBC, beamforming, and several other parameters. For downlink MU packets, there is a lot more signaling as every downlink client can have a different MCS, number of streams, and coding type. To accommodate this extra signaling, MU packets have a variable number of HE-SIG-B symbols. The length and MCS of the HE-SIG-B field is indicated in HE-SIG-A.

12.4 OFDMA

Downlink OFDMA is used in 11ax to minimize the overhead of PHY preambles and MAC backoff when there are a significant number of clients and relatively small packets. The overhead reduction is obtained by combining multiple small packets into one large packet. Without OFDMA, large PHY rates tend to give a high efficiency loss when the packet size per client is relatively small as preamble and backoff times become much larger than the data portion of each packet. The smallest 11ax preamble is 40 μs, so a packet should preferably be at least 400 μs to make the preamble overhead no more than 10%. At a PHY rate of 10 Gbps, however, that means a single packet contains about 500 KB. For a single client, this requires a significant amount of aggregation, which may not be possible if there are latency constraints on parts of the traffic for that client. OFDMA helps in this case as the aggregation can be done over a group of clients.

Figure 12.4 shows the resource unit (RU) allocation for 80 MHz OFDMA. There can be up to 37 simultaneous clients with 26 tones, or 16 clients with

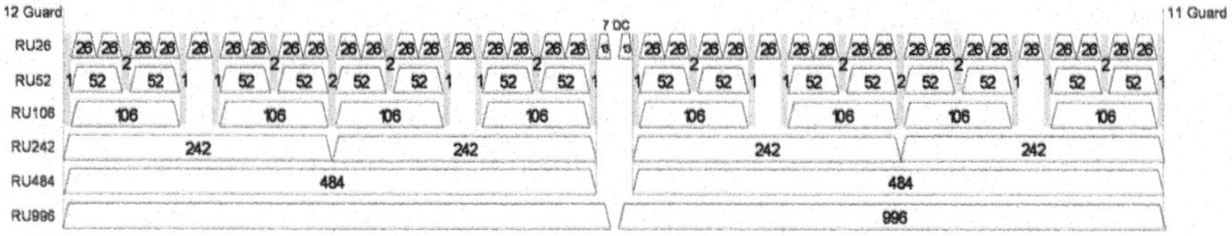

Figure 12.4 80 MHz OFDMA resource unit allocations.

52 tones, 8 clients with 106 tones, 4 clients with 242 tones, 2 clients with 484 tones, or a single client with 996 tones. The same RU sizes are used in other bandwidth modes. In 20 MHz mode, there can be up to 9 simultaneous clients with 26 tones each.

Figure 12.5 shows the packet strucure including signaling per client for an 80 MHz downlink OFDMA packet. The first part of the preamble up to HE-SIG-A is identical for all clients and is repeated in each 20 MHz subchannel to make sure that overlapping networks are able to decode the length of the packet that is encoded in the L-SIG symbol. The HE-SIG-B field is present only in MU packets, either MU-MIMO or OFDMA. It consists of a signaling block that is common to all MU clients and a per-client signaling part. The common part has a separate CRC and tail bits. Each 20 MHz is encoded separately such that a client only needs to decode one 20 MHz subchannel. The per client part has a CRC and tail bits for each group of 2 clients, such that a client only needs to decode a small part of the entire HE-SIG-B correctly in order to decode his signaling info. In case of an odd number of clients, the last group just has one client followed by CRC and tail bits.

Uplink OFDMA is specified in 11ax for the same reason as downlink OFDMA to reduce both PHY and MAC overhead. There is an additional power accumulation gain benefit though; if all uplink clients transmit at the same maximum power in a 26-tone RU than in a full 20 MHz RU in non-OFDMA mode, then the maximum received power at the AP is almost 10 dB more if 9 uplink clients transmit simultaneously in a 20 MHz channel relatively to a single client. For the clients, this means they can transmit at a significantly higher data rate relatively to non-OFDMA mode. This provides a network throughput gain on top of the gain caused by reduced PHY and MAC overhead. Notice that in practice, the power accumulation gain may be limited because of regulatory constraints.

One major difference between uplink and downlink OFDMA is that uplink OFDMA requires time and frequency synchronization of uplink clients, as well as some level of power control. The same is true for uplink MU-MIMO. In an uplink frame, the packets from different clients should arrive at the AP with timing differences that are small relative to the guard interval. To achieve this, clients are required to have an accurate SIFS response time with an error not exceeding $+/-400$ ns. The clients are allowed to transmit uplink frames only in response to a trigger frame by the AP. They have to synchronize their carrier frequency and sampling clock to that of the AP. The accuracy of the frequency synchronization has to better than 1% of a subcarrier spacing to prevent a significant degradation caused by inter-carrier interference.

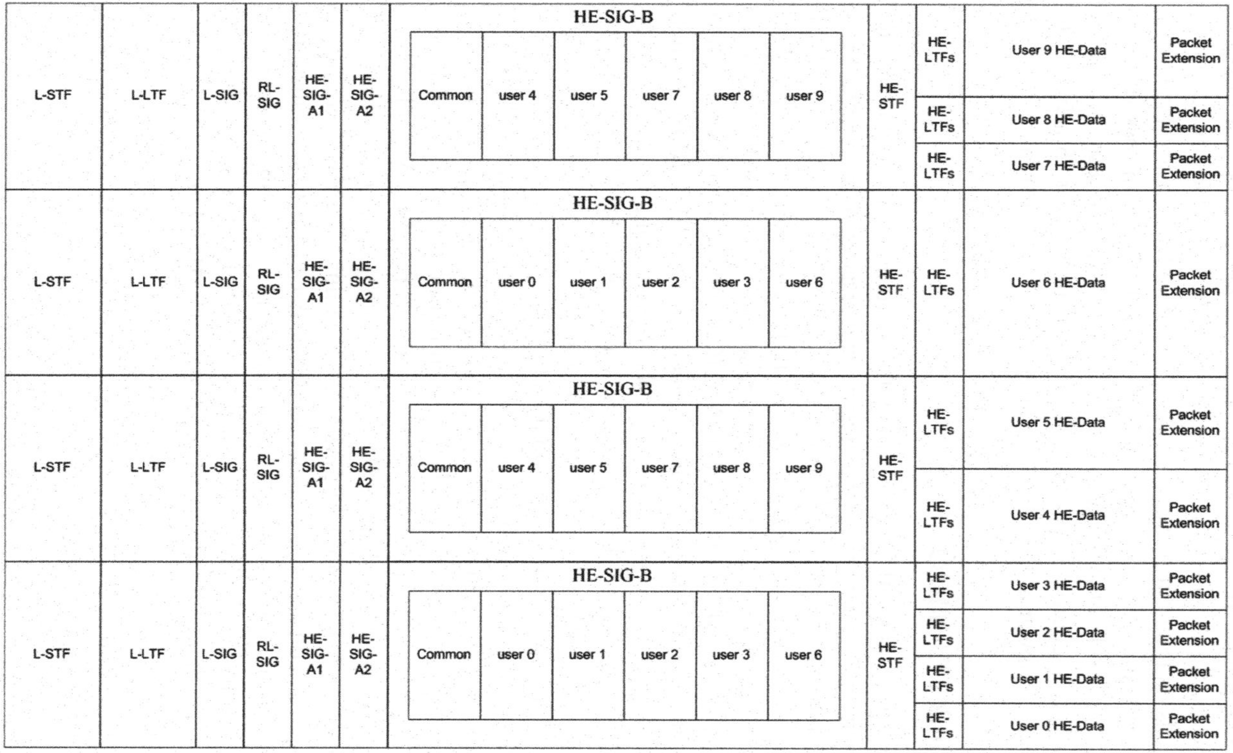

Figure 12.5 11ax packet structure for 80 MHz OFDMA.

Since the subcarrier spacing is 78.125 kHz, the accuracy requirement is about +/−780 Hz. Uplink MU-MIMO and uplink OFDMA clients also have to perform power control with an accuracy of +/−3 dB. To enable power control, the AP signals its transmit power and a target received power level per client in the trigger frame. An uplink client can estimate the path loss from the received power level of the trigger frame and the signaled AP transmit power. By assuming reciprocity, it can then set its transmit power to achieve the desired received power at the AP as requested in the trigger frame.

12.5 Uplink MU-MIMO

In 802.11ac, downlink MU-MIMO was introduced as a way to increase the throughput in cases where the AP had more antennas than the clients. Since then, there has been a demand to increase uplink throughput as well, as the amount of uplink traffic has significantly increased over the past years with smartphone users producing large amounts of video data that is shared across social media. The downlink MU-MIMO throughput advantage is now extended to the uplink in 802.11ax by introducing uplink MU-MIMO. Without MU-MIMO, clients have to compete for the medium and transmit packets sequentially like depicted in Figure 12.6. With uplink MU-MIMO, clients can transmit simultaneously like shown in Figure 12.7, increasing the uplink throughput up to a factor of 4 for 4 uplink clients.

In addition to the throughput multiplication factor, uplink MU-MIMO has the same power accumulation benefit as uplink OFDMA. By having 4 simultaneous uplink MU clients, the received power at the AP is 6 dB more relative to having a single client. This means each uplink MU client can transmit at a higher MCS than in SU mode. Another benefit of uplink MU-MIMO that it benefits downlink MU-MIMO. Downlink MU-MIMO according to 11ac requires compressed channel feedback per client to be transmitted sequentially in the uplink as depicted in Figure 12.8, creating a significant amount of overhead. In 11ax, it is possible – but not required – to transmit the compressed channel feedback in an uplink MU-MIMO mode as shown in

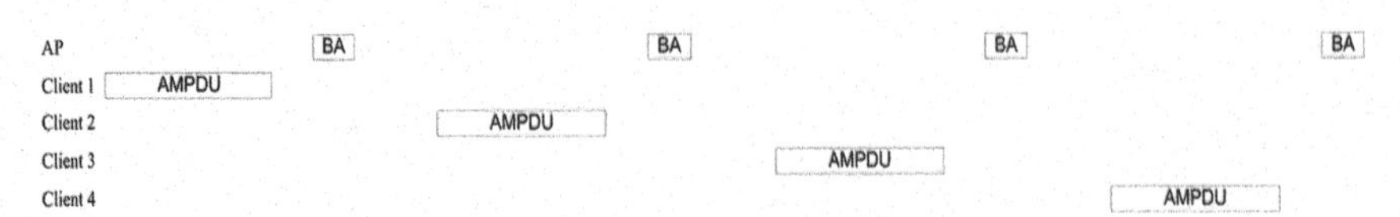

Figure 12.6 Multiple clients sending a packet without use of uplink MU.

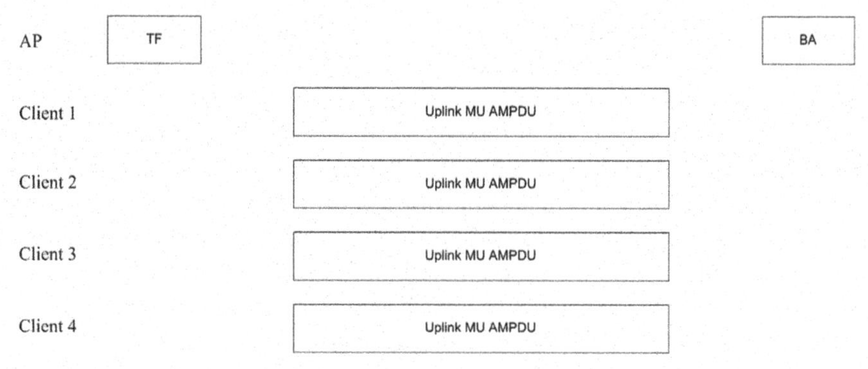

Figure 12.7 Multiple clients sending a packet using 11ax uplink MU.

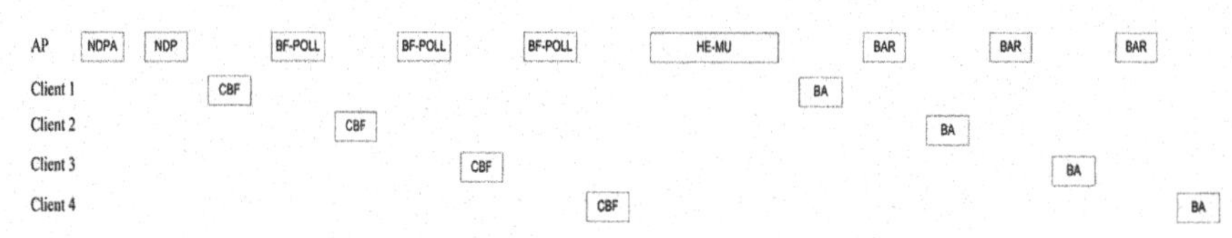

Figure 12.8 Downlink MU TXOP for 11ac.

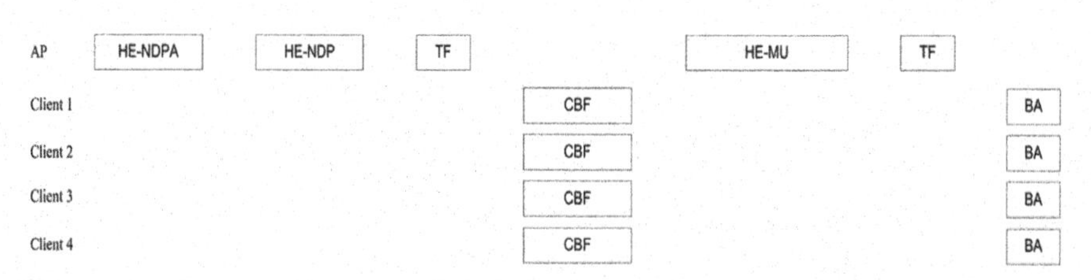

Figure 12.9 Downlink and uplink MU for 11ax.

Figure 12.9. This reduces the overhead and hence improves the throughput of downlink MU-MIMO.

12.6 Range Extension

If current WiFi clients are at the edge of the AP coverage range, they often suffer from the problem that they while they can hear the AP, they cannot associate as the AP cannot hear them. This uplink problem is caused by the fact that the AP generally has more transmit power and hence more range than typical client products. To solve this issue, 11ax defined a range extension mode with a special preamble shown in Figure 12.3. The STF and LTF fields are boosted by 3 dB for the extend range preamble to make sure that the extended range performance is not limited by preamble detection. The preamble signaling symbols are repeated to provide 3 dB of gain in the signaling part. For the data part, Dual Carrier Modulation (DCM) can be used to provide a lowest data rate that is half of the lowest BPSK rate without DCM. DCM basically duplicates half of the tones. To reduce the peak-to-average power ratio caused by the duplication of tone values, for BPSK the odd copied tones are inverted while for QPSK and 16-QAM, the copied tones are conjugated. DCM provides lower rates with twice the diversity order and 3 dB gain relative to no DCM with the same constellation size.

12.7 Dynamic CCA

The current 802.11 standard specifies a few power levels for which any device should defer. There are different power levels depending on bandwidth with higher defer level for larger bandwidth modes. Also, there is a separate level for the case that a valid preamble is detected versus a larger energy detect threshold for cases where the preamble had not been detected, which may happen for instance because the device was receiving some other packet that partially overlapped with another transmission. There are two mechanisms in 802.11ax to improve the defer behavior with the goal of increasing network capacity in dense environments. First, the HE-SIG-A field includes a BSS (Basic Service Set) color which is basically a shortened version of the AP address. After decoding the BSS color, a station knows whether the packet is from its own network or from some overlapping network. If the packet belongs to the same network, it is better to always defer regardless of received power. There is no advantage trying to transmit 2 simultaneous packets in this case because the AP can only receive one packet at a time. It may also be that the received packet is from the AP itself, in which case the AP is not able to receive until it is finished with its transmission. If the BSS color indicates that the packet is from another network, then it may be possible to transmit on top of it. Rather than using a fixed CCA threshold like in current 802.11, 802.11ax allows the use of a dynamic defer threshold where the threshold depends on the transmit power. The lower the transmit power, the higher the defer threshold. The reasoning is that a station with lower transmit power causes interference over a smaller range, hence it can increase its defer threshold such that it defers only for relatively nearby other stations.

12.8 Conclusions

The 802.11ax standard increases the maximum single user and multi-user data rate to about 10 Gbps. Several enhancements are made to increase the throughput in dense networks with many clients, such as the introduction of OFDMA, uplink MU-MIMO, and dynamic CCA. These improvements make 802.11ax attractive for data offloading of crowded LTE networks and also make it a suitable candidate for inclusion in 5G.

References

[1] IEEE P802.11ax/D0.1 draft standard, March 2016.
[2] IEEE Std 802.11ac-2013, amendment to IEEE Std 802.11-2012.

About the Author

Richard van Nee received the M.Sc. degree in Electrical Engineering from Twente University in Enschede, the Netherlands, in 1990, followed by a Ph.D. degree from Delft University of Technology in 1995 (both cum laude). From 1995 to 2000, he worked for Lucent Technologies Bell Labs. In 1996, he developed an OFDM based packet transmission system for the 'Magic WAND' project, which was the first demonstration of OFDM for wireless LAN. Together with NTT, he made an OFDM based proposal that got adopted by the IEEE 802.11a standard. He was also co-author of the CCK-code proposal that got adopted by the IEEE 802.11b standard. The 802.11b CCK codes were originally published in a 1996 paper by Richard van Nee as a way to reduce the peak power in OFDM, but it also turned out to be a useful code set to boost the WLAN data rates to 11 Mbps in 802.11b, while keeping good delay spread robustness and conform to the FCC spreading rules that were still in place at that time. The 802.11b standard brought WLAN data rates on par with existing wired LAN at the time, which contributed to its rapid market adoption and to the formation of the WiFi alliance. In 2001, Richard van Nee co-founded Airgo Networks – acquired by Qualcomm in 2006 – that developed the first MIMO-OFDM modem for wireless LAN and which techniques formed the basis of the IEEE 802.11n standard. He was responsible for the algorithm design of the groundbreaking Airgo technology with many innovations like MIMO-OFDM preamble designs and near-maximum likelihood MIMO decoding techniques. In 2012, he presented the first spec framework of the 802.11ac standard on behalf of a large group of authors. This was the first standard to use MU-MIMO. Together with Ramjee Prasad, he wrote a book on OFDM, entitled 'OFDM for Mobile Multimedia Communications.' This was the first book to describe OFDM for wireless communications, including the first book desciption of the 802.11a standard. It has been used by many engineers worldwide to learn about OFDM for wireless

LAN. He also wrote chapters on wireless LAN in other books, including the Encyclopedia of Telecommunications edited by Proakis, as well as many papers. In 2002, he received the Dutch Veder award for his contributions to standardization of wireless LAN. He holds more than 60 patents related to various WiFi standards and served as an expert witness in several WiFi related lawsuits. He is currently a Senior Director at Qualcomm where he is responsible for WiFi algorithm design and for developing new 802.11 standards.

13

5G for Personalized Health and Ambient Assisted Living

Hermie J. Hermens

University of Twente, Enschede, the Netherlands
Roessingh Research and Development, Enschede, the Netherlands

13.1 Introduction

There is an increasing awareness that healthcare in the western countries cannot be continued in the way it has developed in the past; It is just not sustainable due to the increasing demand for volume of care, limits to the amount of qualified people to work in care and limits to the amount of money we can spend on healthcare. A main trend that is contributing to this need for change is the increased life expectancy of our western society. As the years that are added, come with longer period of time in which we have to cope with chronic conditions, the volume of required health care services is rapidly rising. This increase in volume of care needed cannot be met with a proportional increase in professional personnel and money. There is a strong belief that technology, especially information and communication technology can make a significant contribution to make care more sustainable, by supporting people in their wish for independent living and increase of self management capabilities. Areas that are especially relevant in this context are:

- Unobtrusive monitoring of health related conditions and behavior.
- Home based interventions. Enabling people to do therapy like physical and cognitive training at home, remotely supervised.
- Personalized Health Systems that provide people in real-time personalized advices about their health, based on multimodal sensing.
- Caring homes, supportive environment that intuitively feels the needs of its user and is able to react properly.

These areas are coming up fast due to rapid developments in sensing technology, artificial intelligence and improved communication infrastructures. In this chapter, I would like to describe some recent developments in these areas of health care supporting technology, both from a technical and functional point of view and then, based on the present developments try to imagine how 5G will support and/or accelerate these.

13.2 Technology to Support Personalized Health and Ambient Assisted Living

In this Section, different methods of health monitoring and health assisting systems are discussed. There are diverse projects in all over the world, which they focus on supporting people in need at their home or office without any requirements to visit hospitals or therapists.

13.2.1 Exercising at Home

When patients are able to do their training at home, this would have great benefits for all stakeholders. Patients can do their training where and what time they want, not being obliged to go e.g., three times a week to a physical therapy practice or a rehabilitation centre. If it is partly replacing the intramural therapy sessions, it is also becoming more cost effective and enlarging the capacity of intramural care. If it is used as an extension of intramural rehabilitation, it will help in decreasing the fall back of capacities in the home situation. Figure 13.1 shows variety of activities could be done by remote supervision for patients.

In the EC project Clear, in the period 2011–2013, 1000 patients in 4 European countries were using a system that was able to deliver videos of exercises, an exercise scheme, remote supervision and teleconsultation services. In the extensive health technology assessment it was shown that this remotely supervised treatment was cost effective and efficient when applied as partial replacement of the regular intramural therapy. Moreover, it was shown that the patients, who exercised more, improved more. This underlines that the patient is in the driver seat and is able to take control.

The advantage of this approach is that it is fully web based, making it independent of place and time and relatively low-tech, meaning that you only need a pc or a similar device to get the service working. As such, it is a good first step to introduce this kind of technology in daily healthcare.

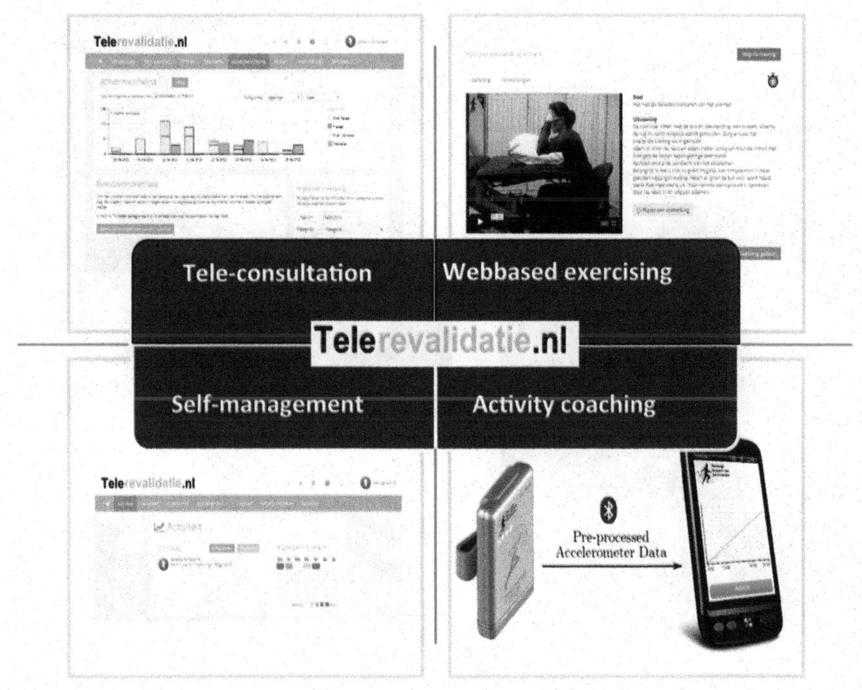

Figure 13.1 Example of a patient platform, supporting web based exercising, video tele-consultation, self management and activity coaching (continuous care and coaching platform, RRD).

A next step in the training at home involves much more complex systems involving games to make the exercises more enjoyable and persuasive, combined with passive or active robotic support systems to correct the movement and/or support the patients in carrying out the desired movement when muscle force is still not sufficient. An example of such a combined system is the Script system, developed in the European Script project; targeting upper extremity training for stroke patients is shown in Figure 13.2. It involves an arm/hand support system that allows personal adjustment of the amount of support, sensors to control the game and a number of games that vary in difficulty and a decision support system that recommends individual game setting and a choice of the games.

With respect to required bandwidth, and technical complexity, systems like Script are much more demanding. Optimally, one would allow experienced clinicians to look at the performance of the patient exercising remotely, but also have real-time control of the robotic device as well as the virtual world where

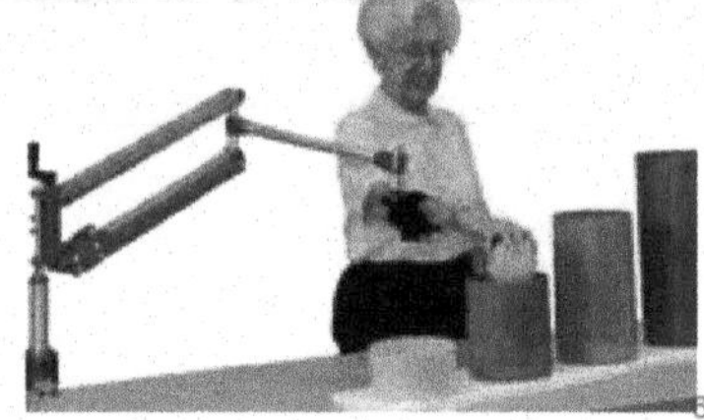

Figure 13.2 Upper extremity exercise system including robotic support for the hand functioning, gravity compensation and individually tailored video games (Scriptproject.eu).

the patient is experiencing. The evaluation studies of the Script system show that such systems are promising with respect to independent and motivating exercising at home, but still rather fragile.

13.2.2 Movement Analysis and Monitoring

Movement analysis is an important step in the functional diagnosis in rehabilitation. It focuses on quantification of the movement pattern and to investigate the underlying deviations in neuromuscular control and the compensation mechanisms that occur.

Gait analysis has been restricted to laboratories, where gait was quantified using optical, marker based systems. Reflecting markers are placed at specific places on the body and assessed using video cameras placed around the subject that pulse with infrared lights to obtain a three dimensional image. Off-line these images are then combined with a three-dimensional model and force plates to calculate joint forces. In the past years, this is being seen more and more as a so-called capacity measurement as the laboratory situation is a rather perfect environment with a very clean context, where the patient can be totally focused on the task at hand. A need has arisen to be able to have quantitative movement data from realistic situations, like at home. In the past years this has been enabled by rapid development of inertial sensor technology and processing algorithms to generate the movements from the data. A recent example of progress made in this area concerns the Interaction project is illustrated in Figure 13.3.

Here a complete movement analysis system has been developed with motion sensors embedded in smart textiles, able to monitor the activities of a person in his own environment. It involves different modalities of sensing like inertial sensors to detect posture and movements, surface EMG sensors to measure muscle activation patterns and 3D force measurement under the foot. The system is presently being applied in stroke patients to monitor how they use their movement capacities in daily life, after a period of rehabilitation. The responses from the clinical users are in general very positive, underlining that these kind of unobtrusive monitoring systems have great promise for future integrated care.

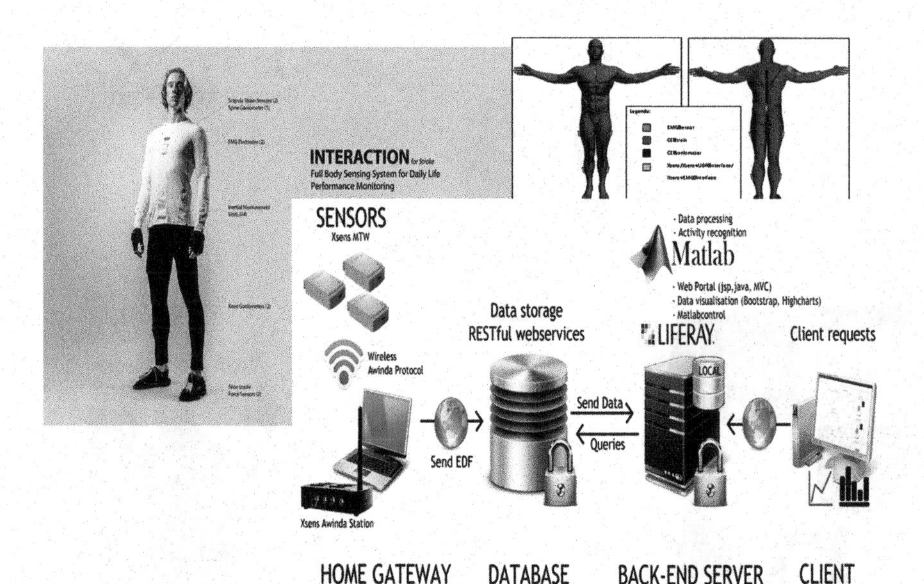

Figure 13.3 The different components of the Interaction monitoring system (https://www.xsens.com/news/interaction-project/).

13.2.3 Personal Coaching Systems

The amount of people with chronic conditions is rapidly growing especially because we live substantially longer but this extension of our lives is going hand in hand with a longer life with chronic conditions. Chronis conditions cannot be cured but it has been shown in many studies that quality of life and the occurrence of co-morbidities is strongly related to the extend we behave healthy, in terms of physical activity, healthy food and avoidance of stress. We, human are not good in changing our behavior, proper and intensive coaching is essential. As we do not have the ability to do this with persons, the idea was born to do this with technical systems using on body and context sensing, smart context aware reasoning and user friendly interaction. The concept of such Personal Coaching System (PCS) is shown in Figure 13.4.

Several PCS have been realized in the past years. Figure 13.5 shows some examples of past and ongoing work. Probably the first PCS was a system to treat neck/shoulder pain during activities of daily living. It involved

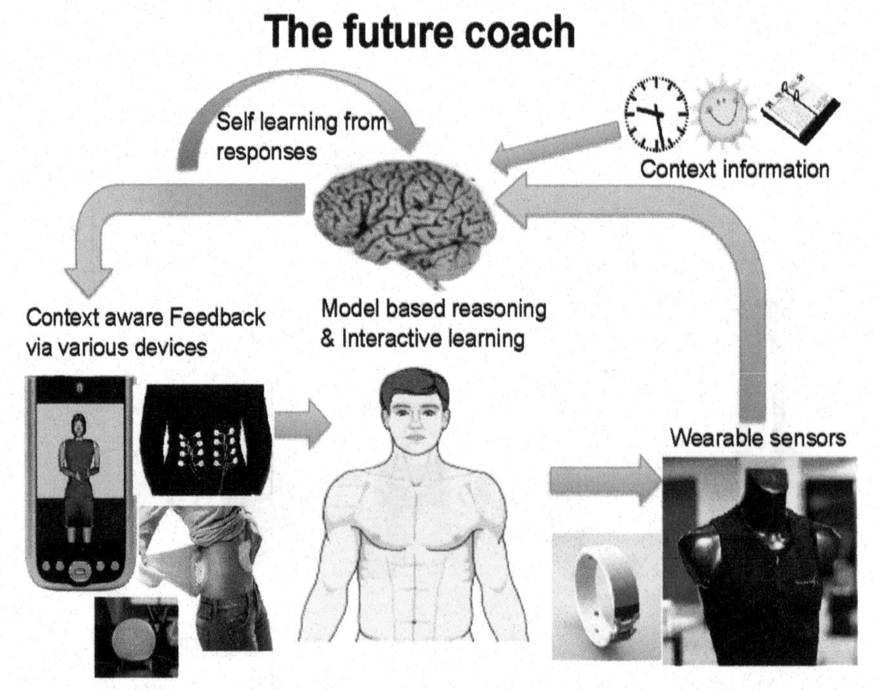

Figure 13.4 Schematic presentation of an artificial Personal coaching system, involving on-body sensing, context aware reasoning and persuasive feedback.

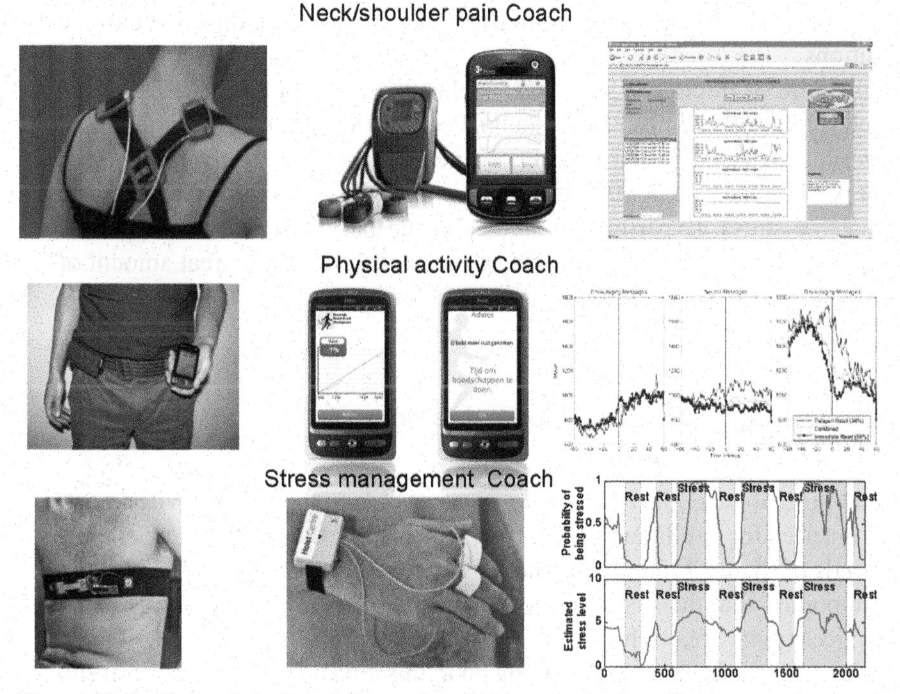

Figure 13.5 Three examples of Personal coaching systems that use sensor and context information to reason about a health condition to provide personalized and persuasive feedback.

measurement of the muscle activity in the Trapezius muscles, well-known to be sensitive for stress and pain, processing the signal in terms of amount of relaxation and giving personal feedback via vibration to warn the subject when there is not sufficient relaxation. This system was successfully evaluated in a European trial, showing the system can be applied in neck/shoulder pain patients, without affecting their activities of daily living.

A second system is a smart activity coach, developed to coach people with chronic conditions towards a more healthy activity pattern through the day. This often concerns a higher level of activity but also in many cases better balancing the amount of activity over the day, to avoid a too early exhaustion. It measures the amount of activity during the day, using a 3D inertial sensor. This measurement is continuously compared with a desired activity pattern and based on the result it provides encouraging, discouraging or neutral messages. Both the content and timing of the messages are dynamically personalized, using a model which takes into account many activity variables and context information. The system is still under development but has been successfully applied in over 100 patients with different chronic conditions.

The third example is a stress management system. Stress appears to be a complex, not well-defined phenomenon. People also respond in different ways to stress, which makes that for a PCS a personal set of sensors are required and a personalized model that is able to weight the different inputs into one estimate of stress. Experimental studies in the lab show that this approach is feasible. Experiments in free living condition show that a great amount of context information has to be taken into account to provide a good real-time stress estimate.

13.2.4 The Caring Home; Supportive Home Environments

In order to enable people to live as long as possible in their own environment, there is a need for intelligent support systems. Ideally, the system should "feel" the needs of the user intuitively and able to react when required. In addition it is important that the system is personalized and able to offer a wide range of services in the area of health, well-being and comfort.

Presently, many projects are going on in this area dealing with several sub goals in the context of the caring home. For example Projects that aims to determine the behavior of the person using different modalities of sensing. Key target population here is people with MCI or dementia as there is a great and rapidly growing need to supervise them as long as they live in their own houses.

A good example of an overall approach to this challenge is the eWall project. The concept of eWall is to offer services in a user-friendly way using a large interactive screen, run in a browser and supported by an extensive cloud service infrastructure. The way the services are presented is innovative by using a 50's home room environment. It offers a personalized set of services including health services like an activity coach, video based exercising, a sleep qualification system and a personalized guidance through the day. Figure 13.6 shows how the eWall screen is look like at patients´ ends.

Integrated approaches, like eWall, require considerable bandwidth for these services as there is lot of sensing, real-time data processing and human computer interaction with animations. It is generally expected that such approach with an integration of services and using a cloud-based infrastructure will become common soon.

Figure 13.6 Example of an integrated caring home environment, offering health, wellness and safety services to people with chronic conditions (http://ewallproject.eu/).

13.3 Opportunities Due to 5G

Overlooking the described upcoming areas that will have impact on the healthcare, especially the care for people with chronic conditions, on the on hand and trying to understand and forecast the contribution 5G can have to these areas, one could make the following predictions.

13.3.1 Movement Analysis and Monitoring

For monitoring purposes, especially in the area of the analysis of functional movements, more and more rather complex multimodal sensors are used, like combinations of gyroscopes, accelerometers, magnetometers and ultrasound to get a complete 3D picture of the subject and his movements.

When these are combined with EMG measurements of several muscles, a considerable bandwidth is required. 5G would allow such simultaneous measurements in a free condition outside the lab. And probably also beyond the present set of measurements, like addition of EEG to assess motor control and mood in real-time would become possible, making a holistic monitoring of all relevant health and behavior aspects possible.

At the other side, the viewing of all these sensor data in an understandable and comprehensive way is a considerable challenge, especially when the data set is expanded extensively. An upcoming technical opportunity is the use of holographic projections. This would allow the presentation of the data to the care professional in a way he/she is used to, like in his daily practice. Seeing the subject move, while information is added like in enhanced reality. The 5G bandwidth would have a big impact in making this feasible.

13.3.2 Personal Coaching Systems

Personal coaching systems are attracting more and more attention as the way forward to support people in changing their behavior. A near real-time response of the PCS is essential, as the user has to be able to link his behavior to the recommendations the system is providing, so he is able to be aware of his behavior and able to change it directly. More complex PCS like one for stress do require an individually modeled combination of several sensors to get an estimate of stress for feedback. The capacities of 5G would support real-time data collection and processing centrally, enabling the use of complex behavior modeling.

13.3.3 The Caring Home

The caring home is just at the beginning of its time. The upcoming Internet of things (IoT) will extend the sensing and interaction capabilities of the home environment considerably and 5G will enable real-time data collection of all the on-body and contextual IoT data, enabling a better intuitive identification of the user needs and required responses of the home environment. It will also enable the incorporation of new sensing methods that involve more data; like mood detection using real-time video processing.

13.4 Discussions

Present healthcare is not yet fully exploiting the capacity of the present 3G and 4G networks. Electronic patient records do use a limited amount of traffic and, apart from the imaging data, involve small amounts of data. Only rehabilitation exercising at home where videos are used to enable independent exercising does require some substantial bandwidth. Nevertheless, it is expected that this will change the upcoming years, now that many health care institutes have their EPD ready and are now looking for technology supported innovations of their care services. And this will mean that the described areas will receive increasing attention. It is expected that 5G will enable significant steps forward in the areas. The expected large bandwidth combined with high reliability and availability, will enable real-time data collection from large amount of sensors (on body, IoT), the use of complex user modeling and new ways of user interaction. It will also increase trust in these areas both by the end-users and care professional community, speeding up the large scale deployment of these care supporting solutions.

References

[1] Amirabdollahian, F. and Ates, S. and Basteris, A. and Cesario, A. and Buurke, J. H. and Hermens, H. J. and Hofs, D. and Johansson, E. and Mountain, G. and Nasr, N. and Nijenhuis, S. M. and Prange, G. B. and Rahman, N. and Sale, P. and Schätzlein, F. and Schooten, B. van and Stienen, A. H. A. (2014) Design, development and deployment of a hand/wrist exoskeleton for home-based rehabilitation after stroke – SCRIPT project. Robotica, 32 (8), pp. 1331–1346. ISSN 0263-5747

[2] Akker, Harm op den and Tabak, Monique and Marin-Perianu, Mihai and Huis in 't Veld, Rianne and Jones, Valerie M. and Hofs, Dennis and Tönis, Thijs M. and Schooten, Boris W. van and Vollenbroek-Hutten, Miriam M. R. and Hermens, Hermie J. (2012) Development and evaluation of a sensor-based system for remote monitoring and treatment of chronic diseases – the continuous care & coaching platform. In: 6th International Symposium on eHealth Services and Technologies, EHST 2012, 3–4 July 2012, Geneva, Switzerland (pp. 19–27).

[3] Akker, Harm op den and Cabrita, Miriam and Akker, Rieks op den and Jones, Valerie M. and Hermens, Hermie J. (2015) Tailored motivational message generation: a model and practical framework for real-time physical activity coaching. Journal of biomedical informatics, 55. pp. 104–115. ISSN 1532-0464

[4] Broens, Tom H. F. and Huis in 't Veld, Rianne M. H. A. and Vollenbroek-Hutten, Miriam M. R. and Hermens, Hermie J. and Halteren, Aart van and Nieuwenhuis, Lambert J. M. (2007) Determinants of successful telemedicine implementations: a literature study. Journal of Telemedicine and Telecare, 13 (6), pp. 303–309. ISSN 1357-633X

[5] Car, Josip and Huckvale, Kit and Hermens, Hermie (2012) Telehealth for long term conditions. BMJ: British Medical Journal, 344 . E4201. ISSN 0959-8138

[6] Hermens, H. and Akker, H. op den and Tabak, M. and Wijsman, J. and Vollenbroek, M. (2014) Personalized Coaching Systems to support healthy behavior in people with chronic conditions. Journal of electromyography and kinesiology, 24(6), pp. 815–826. ISSN 1050-6411

[7] Hermens, Hermie J. and Vollenbroek-Hutten, Miriam M. R. (2008) Towards remote monitoring and remotely supervised training. Journal of Electromyography and Kinesiology, 18 (6), pp. 908–919. ISSN 1050-6411

[8] Huis in 't Veld, M. H. A. and Voerman, G. E. and Hermens, H. J. and Vollenbroek-Hutten, M. M. R. (2007) The receptiveness toward remotely supported myofeedback treatment. Telemedicine and E-health, 13 (3), pp. 293–302. ISSN 1530-5627

[9] Huis in 't Veld, Rianne M. H. A. and Huijgen, Barbara C. H. and Schaake, Leendert and Hermens, Hermie J. and Vollenbroek-Hutten, Miriam M. R. (2008) A staged approach evaluation of remotely supervised myofeedback treatment (RSMT) in women with neck-shoulder pain due to computer work. Telemedicine and E-health, 14 (6), pp. 545–551. ISSN 1530-5627

[10] Huijgen, Barbara C. H. and Vollenbroek-Hutten, Miriam M. R. and Zampolini, Mauro and Opisso, Eloy and Bernabeu, Montse and Nieuwenhoven, Johan van and Ilsbroukx, Stephan and Magni, Riccardo and Giacomozzi, Claudia and Marcellari, Velio and Scattareggia Marchese, Sandro and Hermens, Hermie J. (2008) Feasibility of a home-based telerehabilitation system compared to usual care: arm/hand function in patients with stroke, traumatic brain injury and multiple sclerosis. Journal of Telemedicine and Telecare, 14 (5), pp. 249–256. ISSN 1357-633X

[11] Huis in 't Veld, Rianne M. H. A. and Huijgen, Barbara C. H. and Schaake, Leendert and Hermens, Hermie J. and Vollenbroek-Hutten, Miriam M. R. (2008) A staged approach evaluation of remotely supervised myofeedback treatment (RSMT) in women with neck-shoulder pain due to computer work. Telemedicine and E-health, 14 (6), pp. 545–551. ISSN 1530-5627

[12] Tabak, M. and Vollenbroek-Hutten, M. M. R. and Valk, P. D. van der and Palen, J. A. M. van der and Tonis, T. M. and Hermens, H. J. (2012) Telemonitoring of daily activity and symptom behavior in patients with COPD. International journal of telemedicine and applications, 2012 (438736). ISSN 1687-6415

[13] Tabak, M. and Flierman, I. and Schooten, B. van and Hermens, H. J. (2013) Development of a trusted healthcare service to support self-management and a physically active lifestyle in COPD patients. In: Telemedicine & eHealth 2013: Aging well – how can technology help?, 25–26 November 2013, London, UK.

[14] Voerman, Gerlienke E. and Sandsjö, Leif and Vollenbroek-Hutten, Miriam M. R. and Larsman, Pernilla and Kadefors, Roland and Hermens, Hermie J. (2007) Effects of Ambulant Myofeedback Training and Ergonomic Counselling in Female Computer Workers with Work-Related Neck-Shoulder Complaints: A Randomized Controlled Trial. Journal of Occupational Rehabilitation, 17 (1), pp. 137–152. ISSN 1573-3688

[15] Voerman, Gerlienke E. and Sandsjö, Leif and Vollenbroek-Hutten, Miriam M. R. and Larsman, Pernilla and Kadefors, Roland and Hermens, Hermie J. (2007) Changes in Cognitive-Behavioral Factors and Muscle Activation Patterns after Interventions for Work-Related Neck-Shoulder Complaints: Relations with Discomfort and Disability. Journal of Occupational Rehabilitation, 17 (4), pp. 593–609. ISSN 1053-0487

[16] Vollenbroek-Hutten, M. M. R. and Huis in 't Veld, M. H. A. and Hermens, H. J. (2008) Myotel: adressing motor behavior in neck shoulder pain by assessing and feedback semg in the daily (work) environment. In: International conference on ambulatory monitoring of physical activity and movement, conference book, 21–24 May 2008, Rotterdam, Netherlands (p. 86).

[17] Vollenbroek-Hutten, Miriam M. R. and Hermens, Hermie J. (2010) Remote care nearby. Journal of Telemedicine and Telecare, 16 (6), pp. 294–301. ISSN 1357-633X

[18] Vollenbroek-Hutten, Miriam M. R. and Hermens, Hermie J. and Kadefors, Roland and Danneels, Lieve and Nieuwenhuis, Lambert J. M. and Hasenbring, Monika (2010) Telemedicine services: from idea to implementation. Journal of Telemedicine and Telecare, 16 (6), pp. 291–293. ISSN 1357-633X

[19] Weering, Marit van and Vollenbroek-Hutten, M. M. R. and Kotte, E. M. and Hermens, H. J. (2007) Daily physical activities of patients with chronic pain or fatigue versus asymptomatic controls: a systematic review. Clinical Rehabilitation, 21 (11), pp. 1007–1023. ISSN 0269-2155

[20] Wijsman, Jacqueline and Grundlehner, Bernard and Penders, Julien and Hermens, Hermie (2010) Trapezius Muscle EMG as Predictor of Mental Stress. In: Wireless Health, WH 2010, 5–7 October 2010, San Diego, California (pp. 155–163).

[21] Wijsman, Jacqueline and Grundlehner, Bernard and Liu, Hao and Hermens, Hermie and Penders, Julien (2011) Towards Mental Stress Detection Using Wearable Physiological Sensors. In: 33rd Annual International Conference of the IEEE Engineering in Medicine and Biology Society, EMBC 2011, 30 Aug–3 Sept 2011, Boston, USA (pp. 1798–1801).

About the Author

Hermie J. Hermens did his masters in biomedical engineering at the University of Twente and became head of the research group of the Roessingh Rehabilitation Centre. In 1990 he was co-founder of Roessingh Research and Development, now the largest research institute in the Netherlands in the area of rehabilitation technology and Telemedicine.

He did his Ph.D. in surface Electromyography and became in 2001 Professor in Biomedical Engineering, specialized in human motor control and in 2010 professor in Telemedicine. Currently, he supervises 15 Ph.D. students; over 20 Ph.D. students finished under his (co)supervision.

He is (co)-author of over 300 peer reviewed scientific journal publications and his work was cited over 12000 times (H-index 50). Additional professional functions include: Editor in chief of the Journal of Back and Musculoskeletal Rehabilitation, Past-President and fellow of the int. Society of Electromyography and Kinesiology.

The research focus of Hermens is the creation of innovative health care services by combining biomedical engineering with ICT. He coordinated 3 European projects (Seniam, Crest and Impulse (award successful project)) and participated in over 25 other European projects as key partner. He is strongly involved in many E-Health and Personalised Health related projects, especially at European level (recent: eWall, Perssilaa, In-Life, Deci, Mobiguide).

14

Multi Business Model Innovations in a World of 5G – Towards a World of Advanced Persuasive Business Models Embedded with Sensor- and Persuasive Technologies

Peter Lindgren

Aarhus University, Department of Business Development and Technology, Denmark

14.1 Introduction

Why focus on persuasive business models (PBM) related to sensor- and persuasive technologies in a future world of 5G?

The answer could be

"Technology will always be in need of a Business Model"

Or

"No technology" – neither persuasive technology in a world of 5G – "will go or do without a Business Model" – or in fact many Business Models" (Inspired by Chesbrough) [2, 3]

Further what can the world of 5G provide related to ethics, trust and security of these PBM's.

It appears that researchers and practitioners have yet not researched widely on PBM's related to 5G and how they can and will be applied in the benefit of business, society – and humans. What can businesses and society gain of PBM but what can they probably also loose from the invasion of PBM.

The chapter discusses different aspects and expected evolvement of PBM innovation in a world of 5G in reference of a definition and proposed framework of a PBM.

PBM embedded with sensors and persuasive technology could potentially lead businesses and society into a new area of growth in 2030's 5G world – if 5G technologies and support ethics, trust and security.

Research carried out by Boston Consulting Group back in 2009 showed that "Business Model Innovators earned in an average premium that was four times greater than that enjoyed by product or process Innovators" [12]. Which results can we expect from PBM in a world of 5G?

Businesses focusing on Business model innovation (BMI) have until recently delivered returns that are larger and more sustainable than product and process innovation" [12]. However this is predictably not going to stay on as more and more businesses realize the importance of BMI and gain the competence to do BMI. Further as BM's face the same high speed trends as products did in the early 2000's – the lifetime of BM's shrinks and must even shrink dramatically more every day.

As this race of development and innovation of technologies in the "slip stream" of 5G is going on.

"How will the business models and business model innovation in a world of 5G expectable look like and be influenced by 5G?"

"What will expectable are the next era of BMI related to the evolving 5G technologies developed?"

Well our research in the MBIT group certainly points firstly to the era of multi business models. Secondly to the raise of PBM's embedded with persuasive technologies.

14.2 Persuasive Business Models and Business Model Language

"Technology will always be in need of a Business Model" [2] but Business models will however also in a world of 5G – maybe even more than in the past – be in need, based on and dependent on advanced technologies.

BMI these days is influenced by the growing numbers of sensors embedded in everything, everybody and anywhere. Just in 2020, 300 billion sensors are expected spread out in our society – making our lifestyle and thereby BMI different. The intelligent sensors and not least printed, flexible and biological sensors are expected to change the game of BMI. Sensor technology will e.g., mean inspired by Bryzek [1] and Drupa [4]:

- **'unobservable' sensing BM** which will break new ground in the technology **component** of the BM e.g., the dimension of sensing biohazards, smells, material stresses, pathogens, level of corrosion and chemicals in BM's.
- **Micro-sensor implants** in humans, animals and environments – which will track e.g., the healing process for internal injuries, illness, enable

health and animal care professionals and machines to take remedial action based on continual data – big and mega data – from the BM eco system.

- **Biodegradable sensors** monitoring e.g., soil moisture or nutrient content for optimum crop production or healthcare for human's suffering from diabetes.
- **Self-powered sensors** powered by using the heat difference between the human's, animal's or machines "body" and surrounding air.
- **Self-healing sensors** repairing themselves in the event of disaster or other structural disruptions.
- **Live cell-based sensing** e.g., an amalgamation of sensor technology and living cells, allow scientist and business people to understand the biological effect of medicines, drugs, environment and biohazards.
- **Sensor swarms** coordinate their activities, deciding interactively what to measure and where through a self-learning system directing their movements, data collection and persuasiveness.
- **Smart dust**, microscopic sensors powered by vibrations, monitor situations ranging from battlefield activities, structural strength of buildings, clogged arteries or BM ecosystems.

The sensor technology in 2030 will expectable be even more advanced than today – as expected covering all 5 sensors – and enabling all 5 sensor technologies to speak together and be intelligent integrated [5–7, 11].

We thereby step into the intersection of different sensors, industries, disciplines, business models, cultures and backgrounds – in other words, embrace diversity in technologies and business models – of thought, perspectives, experience, expertise – and values – which will drive BMI into a new era.

Inspired by the Medici family in Florence, whose patronage of artists, architects, scientists, philosophers – the interdisciplinary approach – which in the Renaissance helped bring about a new age of creativity, discovery and innovation in Europe – is now turning back – The concept of the Medici Effect [9].

An increasing development of persuasive technology enabled by the evolvement of sensor technology, big data and 5G changes the game of BMI. The evolution and era of persuasive business models will result in more advanced PT's expected to come, evolve with tremendous speed and having such impact on business, society and humans never seen and imagined before.

PT [15 A] is still a vibrant interdisciplinary research field, focusing on the design, development and evaluation of interactive technologies – changing users' – human and machines – attitudes or behaviors through persuasion and social influence. PB has still to be accepted and be researched deeply. Persuasive technologies can with the aid of 5G be used to change peoples and machines behavior in various domains. Persuasive technologies will be a vital component in the competence dimension of any persuasive business model in the future [16]. Expectably product-, service-, production- and process technology will be embedded with persuasive technologies in future business models – and thereby become PB BM's:

"the persuasive business model are strategically designed with the aim to change users, customers, networkpartners and employees behavior via its value proposition(s) acting interactively together with the other 6 BM dimensions and related BM's" [16].

The application of the persuasive BMs will expectably really take off with "the 5G roll-out" in the next 5 to 10 years [15]. The development of sensors and persuasive technologies makes it possible to create business model ecosystems with persuasive business models. Although persuasive business model use persuasive technologies and are basically created to "persuade" for a certain behavior in accordance with the strategy of the BM and business numerous examples shows that they have not just advantaged but also a backside. They should therefore be ethically being constructed secure – but today however they are mostly uncontrolled, unregistered and free of use.

We claim – there are some steps to be taken before we – the businesses and society reach a deeper understanding of how persuasive businesses really can use persuasive business models and what they really can do with persuasive business models.

On behalf of inputs from SW2010 [18]–SW2015 [19], lab experiments in the MBIT and Stanford Peace Innovation Lab together with state of the art persuasive business model and technology research we conceptual elaborate on a futuristic outlook to persuasive business models. What can we expect of persuasive business Models and persuasive business model innovation in a future world of 5G.

14.3 A World of 5G and Persuasive Business Models

"In the past ten years the number of sensor-, wireless and persuasive technologies in our everyday life, have increased many-fold. We are now moving fast towards a world of 5G which obviously will by standard have embedded persuasive technologies [4, 5, 15, 17] – and therefore it will soon be a reality to business to deal with these persuasive technologies. "The study of these persuasive technologies, and how they affect our lives and routines are still very young – and we know little about how they will affect our future lives – but we know that what we can expect of persuasive technologies cannot even we imagine today [6, 15, 17].

Researchers, business and public players alike are keenly devoting themselves to understanding how these different persuasive technologies might be designed, so that desirable technologies, behaviour and not least "business models" are obtained and can be created, captured, delivered, received and consumed – hopefully secure and sustainable. There is large expectation to that 5G and persuasive business models both will enable better business. Especially healthcare and well care sector expects much of persuasive technologies [13] to overcome some of their big economic burden due to amongst others a growing elderly population and increasing medicine costs.

The power and importance of persuasive technologies embedded in a multitude of business models is therefore obvious! – and to some extent would some say – on the dark site scary – if not secured and lead in a valuable and sustainable directions. However the evolvement of persuasive technologies and persuasive business models are not to be stopped and their impact will be enormous in the future and even be clearer – as we move into a world of 5G.

14.4 Persuasive Business Models and Business Model Language in a World of 5G

Today most businesses are not really able to carry out and innovate persuasive business models. As the sensor-, persuasive- and 5G technologies develop there is suddenly no excuse and barriers technologically to not innovate persuasive business models. The challenges lay a whole different area. There is not yet an alignment and an accepted business model language in the business model community [20–23] are one of the major reason to this "barrier" – preventing business to take the next step into persuasive business modeling. A common agreed business model language is highly needed to make it possible for business to communicate their Business Model dimensions and components with one another – but more important using the full potential of persuasive and sensing business models [13, 14]. Many businesses are still what could be classified as "Business Model analphabetic".

In a world of 5G this language will be even more important to develop and have clarified. If all businesses could agree upon a common business model language then persuasive business modeling and the use of persuasive business models could really take off – and not just be reserved for the use of large businesses with powerful computers and high skilled employees.

Today most businesses are just able to see a 2D mapping of their BM value exchange as we illustrate from one of our research projects in Figure 14.1.

5G will enable them with technologically to see data and get information of much more of the BM and its value stream. The abilities to really "see" the complex world of Business Models and the impact of Business Model Innovation.

Due to all the interactions that 5G technology will give us combined with advanced visualization technology business will be able to "see" and sense much more – and faster. It will provide them with enormous amount of new data, knowledge and insights, which they can be able to understand when they agreed on "a common language" and where to look [10].

"Unwrapping" BM knowledge in small lab experiments with a small part of a business – in this example just with few different BM's the picture shows us a rather complex and less operational picture of the value exchange between business models. This is shown in Figure 14.2.

Understanding this complex Business Model value creation, capturing, delivering, receiving and consumption process, of both tangible (full line) and intangible (dotted line) value exchanges between Business Models will however be essential and form the platform for persuasive business modeling in a world of 5G.

Our hypothesis is that PBM's will be in common in near future but before that we will see several generations of PBM. We expect that the real application

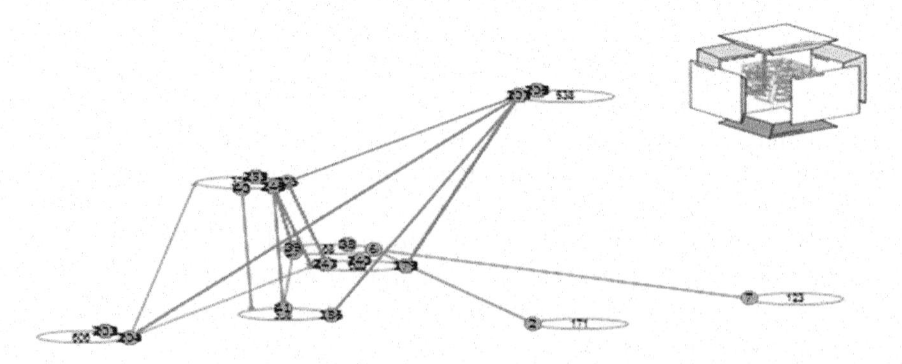

Figure 14.1 2D Mapping of one BM value exchange sequence from an industrial business [16].

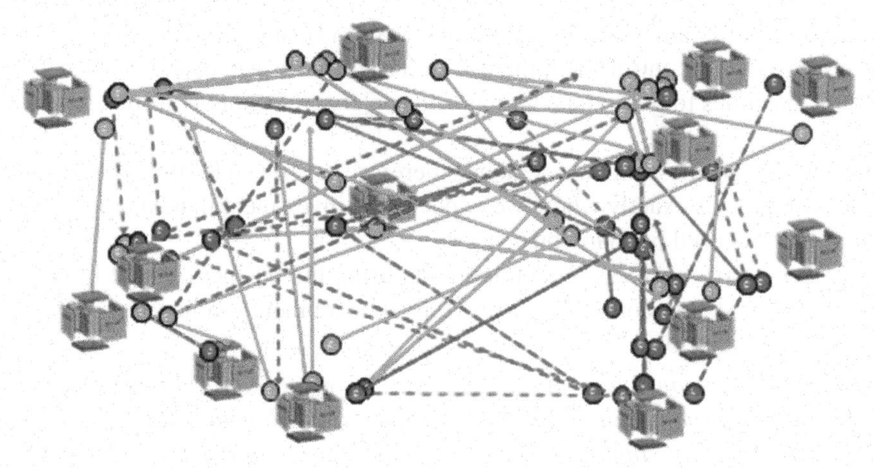

Figure 14.2 2D Mapping of a multitude of BM value exchange sequence between BM's inside a business [16].

of the persuasive BMs would begin with the 5G roll-out in the next 5 to 10 years.

Both users, customers, networkpartners, employees and businesses of today – in general will find initially those persuasive business models and persuasive business model innovation processes highly risky, foreign and radical related to existing BM's and BM innovation processes – that they've been experimenting with in their past. Certainly society and politicians need soon – and fast – involve themselves in the debate and rulemaking of Persuasive Business Modeling.

The history of the persuasive business model concept is still relatively young. As 3G and 4G based business ecosystems emerged, many business (Google, Face book, Amazon, Ebay, Zinga, Blizzard began rethinking their business model and business model structure [16]. They began to build in persuasive components (motivating colors, text, tabs, sounds), dimensions (value propositions, value chain activities, relations and networks) even BM's which could "motivate" – some would say "persuade" – users, customers, network and employees to certain behaviors. These attempts were initially rather simple and have up to for some years ago been very harmless and relative simple constructed.

However when highly professional promotion experts, psychologists, sociologist, computer scientist and business model experts are brought together in interdisciplinary BMI teams with the aim of creating persuasive business models – then the next generation persuasive business model innovation could really begin and find their way to Business Model Ecosystems (BMES). If no control and influence – then situations where people, business and machines forget to take human and society ethics into consideration begin to emerge.

Persuasive business models demands therefore several and different competences and this is exactly where we expect BMI investment in next coming years will be focused by many businesses.

The persuasive business model approach bring the business model in to a more advanced step compared to previous development – a new era – by answering the question –

What if we really could use the sensors and the mega data they generate and could create knowledge in real-time interaction with people, things and businesses – to really "influence" or "persuade" for certain behaviors?

It proposes that the persuasive business model in a world of 5G becomes a business model framework that is reasonably simple, logical, measurable, comprehensive, operational and meaningful. The persuasive Business Model we propose as related to 7 dimensions can be seen in Table 14.1 [16].

Our previous research [13] show that a persuasive business model at an optimum adapt a multi business model approach combining and relating

Table 14.1 Generic dimensions and questions to any persuasive business model [16]

Dimensions in a Generic BM	Core Questions Related to a Generic Persuasive BM
Value proposition/s (products, services and processes) that the business offers (Physical, Digital, Virtual)	What are our value propositions?
Customer/s and Users (Target users, customers, market segments that the business serves – geographies, physical, digital, virtual).	Who do we serve?
Value chain [internal] configuration.(physical, digital, virtual)	What value chain functions do we provide?
Competences (assets, processes and activities) that translate business' inputs into value for customers and/or users (outputs). (Physical, digital, Virtual)	What are our competences?
Network – Network and Network partners (strategic partners, suppliers and others (Physical, digital, virtual)	What are our networks?
Relations(s) e.g., physical, digital and virtual relations, personal. (Physical, digital, virtual)	What are our relations?
Value formula (Profit formulae and other value formulae. (physical, digital, virtual)	What are our value formulae?

different "ingredients" from more than one business model, meaning that persuasive business models are in relations with other persuasive business models and they have embedded, interactive, dynamic persuasive technology built in.

It can be argued that strategy [16], are simply embedded with the persuasive business model approach, providing the larger platform for the business who strategically have decided to be based upon and act with one or more persuasive business models.

With other words – a persuasive business model includes an interactive, dynamic business model and business model innovation strategy vision, mission and goal(s) where the PBM seeks to achieve impact on business models including users, customers, and technologies – all dimensions of other business models [16].

The persuasive business model will in this context by nature try to attach to anything, anybody, anywhere and anytime – with the overall aim to "persuade". 5G is expected to be the backbone of these persuasive business models and persuasive business model ecosystems.

5G enables both the vision of creating persuasive business models any time, any place, with anybody and anything. In the future all human beings, all things at any time and in any place will have the possibility to act persuasively.

We expect that the persuasive business model based on advanced sensor- and persuasive technology will indeed not only be important but also add increasingly high value to the stakeholders involved. However we still have some challenge to overcome before we reach a final destination where all business models are persuasive in a world of 5G – hereunder hopefully respecting and acknowledge security, trust and ethic challenges.

An even stronger focus on security, personal security, and network based security technology must be expected of a 5G world. Business models that are continuously persuasive, in process and changing – run by businesses, humans but importantly more and more machines – continuously in different BMES context sets 5G business and researchers under high pressure to quickly find solutions – both technological and business wise to meet the requirements of all kind of stakeholders for increasing agility, flexibility, individualization, privacy related to persuasiveness.

A concept proposal for persuasive technology and business models which are independent of time, place, bodies and things – and at the same time are secure was proposed earlier (Lindgren 2016) as seen in Figure 14.3. This was however shown as proposed in a first generation proposal – an "ecosystem" of secure persuasive business model innovation showing the future context

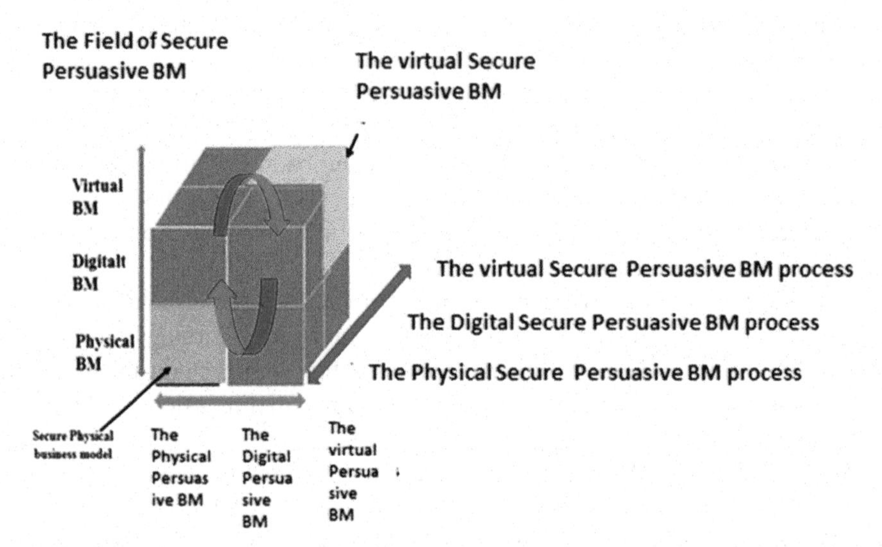

Figure 14.3 A "ecosystem" of the secure persuasive physical, digital and virtual business model [16].

to come – the full integration between physical, digital and virtual persuasive business models operating in a secure context [16].

The fulfillment of embedding persuasive BM's anytime, anyplace, with anybody and any things seems to be achievable in the very near future of 5G. Persuasive business models in a world of 5G will not only be a matter of security on the "surface" of things, places, people, animals and time but also will be about persuasive business models placed both inside and outside things, bodies, species and time.

14.5 Conclusions

Persuasive business models in a world of 5G we expect will become one of the most important BMI concepts for future business and business model innovation. It will also be one of the most discussed topics related to ethics, trust a security – actually the discussion has already begun [16].

Persuasive Business Models are integrated and embedded with advanced persuasive business model technology. It is in this context that the fast evolvement of persuasive business models related to the vision of the secure, persuasive and sustainable business; business model ecosystem, business model ecosystems and world of 5G should be seen.

The secure persuasive business model concept has not yet been fully realized – but business and societies in general continue to evolve and embrace new perceptions of persuasive business models as the challenges of business model innovation gets more complex, their business models lifetime shrinks and the opportunities of the persuasive business model growths with the emerging world of 5G.

Persuasive business models will exist in physical, digital and virtual worlds – operating in a continuous process – integrated, agile, dynamic and better connected business model ecosystems. Delivering in a continuously process of persuasive value propositions – wherever, whenever, whatever the user, customer, network partner, employees, things and business demands it. Persuasive business models will at an optimum operate together in a multi business model setup – a business model network collaboration outside and inside things, bodies, species and businesses in the future.

One must imagine that all 7 dimensions of a persuasive BM can and will change continually in a world of 5G and persuasive business model innovation processes. This makes it extremely difficult to measure, control persuasive BM and BMI processes and thereby control and leadership from outside either it is a user, customer, network partner, competitor or society is highly necessary to begin.

Expectations are that majority of all future business models will be persuasive. This of course only if human being, global society and the business behind the business models including the technology will allow it.

Future persuasive business models will use and take advantage of all opportunities of the 5G technologies given. How persuasive business models can be controlled in this context is however still for the 5G community to discuss and research.

References

[1] Bryzek, J Fairchild (2013) *"Emergence of Trillion Sensor Opportunity," SemiconWest, http://www.semiconwest.org/sites/semiconwest.org/files/docs/SW2013_Janusz Bryzek_Fairchild Semiconductor.pdf.

[2] Chesbrough, H. and Rosenbloom, R. S. (2000) The role of the business model in capturing value from innovation: Evidence from XEROX Corporation's technology spinoff companies, Industrial and Corporate Change, Vol. 11, No. 3, pp. 529–555.

[3] Chesbrough, H. (2006) Open Business Models. How to Thrive in the New Innovation Landscape, Harvard Business School Press.

[4] Drupa (2013) ***"IDTechEx: Printed sensors market will increase by more than $1 billion by 2020," Drupa, http://www.drupa.com/cipp/md_drupa/custom/pub/content,oid,30443/lang,2/ticket,g_u_e_s_t/local_lang,2
[5] Fogg, B. J. (2003). Persuasive technology: Using computers to change what we think and do. San Francisco, CA, USA: Morgan Kaufmann Publishers.
[6] Fogg, B. J., & Eckles, D. (Eds.). (2007). Mobile Persuasion: 20 Perspectives on the Future of Behavior Change. Stanford, California: Stanford Captology Media.
[7] Fogg, B. J. 2012. Persuasive Technology Stanford University Press Persuasive Technologies.
[8] Intille Stephen S. (2004). A New Research Challenge: Persuasive Technology to Motivate Healthy Aging IEEE Transactions on Information Technology in Biomedicine, Vol. 8, No. 3, September 2004, 235.
[9] Johansson Frans (2004), The concept of the Medici effect – Breakthrough Insights at the Intersection of Ideas, Concepts, and Cultures Harvard Business School Press. ISBN 1591391865
[10] Horn Ole Rasmussen et al. (2016) How important are business model relations in innovation management? Preceedings – ISPIM Innovation Forum forum.ispim.org Boston
[11] Ligthart, L. P. and Ramjee Prasad (2016) "The Role of ICT for Multi-Diciplinary Applications in 2030 – Conasense River Publisher.
[12] Lindgarth, Z et al. (2009) "Business Model innovation – When the game gets tough – change the game" Research carried out by Boston Consulting Group.
[13] Lindgren P., Morten Karnøe Søndergaard, Mark Nelson, and B. J. Fogg (2013). Persuasive Business Models Journal of Multi Business Model Innovation and Technology, Vol. 1, 70–98. River Publishers.
[14] Lindgren P, and Annabeth Aagaard (2014). The Sensing Business Model Wireless Personal Communications May 2014, Volume 76, Issue 2, pp. 291–309.
[15] Persuasive technology Conference Austria 2016. http://persuasive2016.org/
[16] Prasad Ramjee and Dr. Sudhir Dixit (2016) Beyond 2050 The Wireless World in 2050 and Beyond: A Window into the Future River Publisher – Chapter xx
[17] Prasad R. (2014) "5G and Beyond" Springer Verlag.

[18] SW2010 Strategic Workshop 2010 held at Hotel Relais Certosa, Florence, Italy MAY 26–28 , 2010 – 12TH STRATEGIC WORKSHOP 2010 Distributed and Secure Cloud Clustering (DISC).

[19] SW2015 Strategic Workshop 2015 held at Villa Mondragone Via Frascati, 51, 00040 Monte Porzio Catone, Napoli Italy Human Bond Communications (HBC) Seventeenth Strategic Workshop (SW'15) May 18–20, 2015.

[20] Teece, D. J. (2010). Business models, business strategy and innovation. Long Range Planning, 45(2–3), 172–194.

[21] Teece, D.J. 2015, "Chapter 16 – Technological Innovation and the Theory of the Firm: The Role of Enterprise-Level Knowledge, Complementarities, and (Dynamic) Capabilities" in Handbook of the Economics of Innovation North-Holland, pp. 679–730.

[22] Zott, C., & Amit, R. & Massa, L. 2011 The business model: recent developments and future research "Electronic copy available at: http: //ssrn.com/abstract=1674384"

[23] Zott, C. & Amit, R. 2013, "The business model: A theoretically anchored robust construct for strategic analysis", Strategic Organization, vol. 11, no. 4, pp. 403.

About the Author

Peter Lindgren, holds a full Professorship in Multi business model and Technology innovation at Aarhus University – Business development and technology innovation and has researched and worked with network based high speed innovation since 2000. He is author to several articles and books about multi business model innovation in networks and Emerging Business Models. He has been researcher at Politechnico di Milano in Italy (2002/03) and Stanford University, USA (2010/11) and has in the time period 2007–2010 been the founder and Center Manager of International Center for Innovation www.ici.aau.dk at Aalborg University. He works today as researcher in many different multi business model and technology innovations projects and knowledge networks among others E100 – http://www.entovation.com/kleadmap/, Stanford University project Peace Innovation Lab http://captology.stanford.edu/projects/peace-innovation.html, The Nordic Women in business project – www.womeninbusiness.dk/, The Center for TeleInFrastruktur (CTIF) at Aalborg University www.ctif.aau.dk, EU FP7 project about "multi business model innovation in the clouds" – www.Neffics.eu. He is co-author to several books. He has an entrepreneurial and interdisciplinary approach to research and has initiated several Danish and International research programmes. He is one of the founders of MBIT research group and the CTIF Global Capsule.

His research interests are multi business model and technology innovation in networks, multi business model typologies and new global business models.

Index

Lightning Source UK Ltd.
Milton Keynes UK
UKOW06n1147270117
293027UK00002B/26/P